ELEMENTARY STATISTICS:

A CONCEPTUAL APPROACH USING A SPREADSHEET

Bayard Baylis
Cornerstone University

Carlos A. Pereira
Roberts Wesleyan College

Barbara Rose
Roberts Wesleyan College

Elementary Statistics:
A Conceptual Approach Using a Spreadsheet

ISBN 0-9666708-0-9

Cover Design: *Keith DeWaters*
Printer/Binder: *EDMAP*
296 Harper Street, Nelsonville, OH 45764
740-753-3439
www.edmap.biz

A WORD TO THE STUDENT

H. G. Wells once remarked, "Statistical thinking will one day be as necessary for efficient citizenship as the ability to read and write." You may not necessarily want to accept his conclusion, but today's society is much more dependent upon statistics and statistical reasoning than society when H. G. Wells made the statement. You need only watch one hour of television news to see the evidence: whether it is a discussion of the unemployment rate in our cities, or the GNP of developing nations, or the fatality count and damage estimates from the latest natural disaster, or the toothpaste commercial between news reports.

Why are you studying statistics? For many of you it is a requirement within your program. For others, statistics may have been suggested as an area with which you should be familiar. For some of you, statistics may be a curiosity. For some, there may be an intense desire to master the subject. For others, statistics may be a dreaded experience.

Statistics, like most subject areas, can be taught and studied at several different levels. Each of you will be bringing a different set of experiences and expectations to this course. Some of you have had several mathematics or other quantitatively oriented courses. For others, this will represent your first quantitatively oriented college course.

In the face of all this diversity, what do we, as authors, expect you to accomplish with the help of this text, and how do we expect you to do it? The main "academic" objectives of this text include an appreciation of the usefulness of the statistical method in your professional field, an understanding of the assumptions and logic underlying the application of statistical tools, the ability to select the appropriate statistical techniques and perform the necessary computations, and the ability to interpret and understand your own results and the results of others.

The advent of the pocket calculator and the personal computer has brought a new dimension to the teaching and learning of statistics. No longer does the student (and the instructor) have to worry about adding long strings of numbers, and then laboriously calculating a square root. The instructor and student may now concentrate on the process. In doing this, more relevant and more "real" problems can be treated in an introductory course; the student can see statistics in action. The emphais of this book is not upon calculations. Although you may want to have a calculator with statistical functions available to do some of the mathematics by hand, this book assumes the reader has access to a computer that can perform the various statistical procedures introduced in the following chapters. Many of the formulas will be listed at the end of the chapters for reference or for you to use them if you desire, but the intent is that most of the "statistics" will be done on your computer. Tables for the distributions introduced in the book will also be included in the appendices for those who might want to reference them.

You would be deluded to think that you could label yourself a "statistician" after completing this text (or any similar text). That only comes after years of "living with it." This text can at best be an introduction. But it is a necessary introduction, since many disciplines rely on statistical tools and techniques for much of the ongoing research.

Besides professional objectives, there are also personal objectives for this text. These include the habit of applying statistical concepts and analysis to daily activities and the development of a healthy attitude of skepticism with respect to statistical information. To be an effective adult citizen, you should not be afraid to use statistics when appropriate. Nor, should you be afraid to question statistics that appear to be used out of their proper context.

These professional and personal objectives are fine. However, every other introductory statistical text will emphasize these same objectives. Why should we expend the effort to write another one and why should you expend the effort to study this one? Because we believe there is one more objective that must be achieved by every educated adult -- the ability to ask the right questions.

In the "real" world, you usually do not come across situations that directly ask you to calculate the chi-square value of a set of data. However, there are many situations in which you may be called upon to judge whether some data "fits with what was expected." For example, suppose your company is involved in the manufacture of facial soap. The process you use to make the soap produces a soap that is naturally "muddy brown" in color. Will the color make any difference to your potential customers? Or, should you introduce one more step in the manufacturing process, with the associated costs, to change the color to "something more attractive" to customers. Intuition tells you that consumers will not buy "muddy brown" facial soap. How can you know whether to go to the extra expense to change the color? One way is to run a marketing test. You provide 100 homes with samples of three "different" facial soaps that you are "considering." Although you imply that there are significant differences in the soaps, the only real difference is the color. You leave one the natural "muddy brown." You color one white, and the third a pastel pink. You ask families to use the three soaps for several weeks. After three weeks you call the families, ask them how the testing is progressing, and ask them which of the soaps they prefer and why. You tabulate the results of the poll and ask "So what?" This problem is an excellent example of a situation in which the "chi-square goodness-of-fit" test should be used.

This example illustrates the process of "asking the right questions." "What do I know? What do I want to know? How do I get from what I know to what I want to know?" Although this process is easily stated, it is not always easily applied. What we have tried to do in our teaching, and in this text, is to provide students with a guide to the art of questioning. We do not view this text as a "cookbook" with a number of recipes. Rather, it is more like a road map leading to the desired destinations.

We have attempted to model this process throughout the text. New material will be introduced by looking a situation, and asking: "What do we know? What do we want to know? How do we get from what we know to what we want to know?" Much of the material used in this text will be "real" data, or data that looks real. We will use actual problems from the disciplines of management science, marketing, psychology, sociology, biology, medicine and others. We will also invent problems and data that look very much like the problems and data from real situations. If a problem is a real problem, it will be so noted.

We hope that this text will help you learn statistics. But we also hope that this text will help you learn a skill which is far more important than statistical reasoning. Asking the right questions may be the most important skill you can develop as an educated person.

Bayard Baylis
Cornerstone University
Grand Rapids, Michigan

Carlos A. Pereira
Roberts Wesleyan College
Rochester, New York

Barbara Rose
Roberts Wesleyan College
Rochester, New York

ACKNOWLEDGMENTS

We thank the classes of Dr. Rose, Dr. Pereira, and Karen Scheske for all of their notes, questions and suggestions.

We especially thank our families for putting up with many hours, many trips, and the many times we used the excuse, "I've got to work on the book."

Dedicated to our precious gifts from God.
Our grandchildren

Katelynn Ann Reed
Chase Tooms
Cole Jayden Tooms

Elise Reneé Pereira
Luc Nathanael Pereira

Josiah Aaron Ottolini
Micah Lee Ottolini
Hanna Rose Ottolini

TABLE OF CONTENTS

A Word To The Student i
Table of Contents iv

Chapter 1 INTRODUCTION
1.1 Introduction 1
1.2 Definitions 4
1.3 Research Design 7
1.4 Descriptive Statistics 8
1.5 Inferential Statistics 10
1.6 Ethical Considerations 10
1.7 Summary 14

Chapter 2 RESEARCH DESIGN AND DATA COLLECTION
2.1 Introduction 15
2.2 Discrete Data 16
2.3 Continuous Data 18
2.4 Data Collection and Research Design 21
2.5 Samples and Sampling 24
2.6 Summary 30

Chapter 3 DATA PRESENTATION
3.1 Introduction 31
3.2 Frequency Distributions 31
3.3 Graphs 37
3.4 Ethics of Graphs 42
3.5 Summary 45

Chapter 4 DESCRIPTIVE STATISTICS
4.1 Introduction 46
4.2 Measures of Central Tendency 47
4.3 Measures of Dispersion 54
4.4 Measures of Location 59
4.5 Exploratory Data Analysis 66
4.6 Summary 69

Chapter 5 PROBABILITY AND SAMPLING DISTRIBUTIONS
5.1 Introduction 71
5.2 Probability 72
5.3 Random Variables 79
5.4 Continuous Distributions 83
5.5 Normal Distributions 87

5.6 Central Limit Theorem 90
5.7 Point Estimates 98
5.8 Interval Estimates 101
5.9 Summary 107

Chapter 6 HYPOTHESIS TESTING
6.1 Introduction 109
6.2 Types of Hypotheses 109
6.3 Types of Errors and Level of Significance 113
6.4 Large Sample Test of the Mean 117
6.5 Small Sample Test of the Mean 122
6.6 Summary 125

Chapter 7 HYPOTHESIS TESTING FOR TWO POPULATIONS
7.1 Introduction 127
7.2 Two-Sample Designs 127
7.3 Difference of Means (dependent samples) 129
7.4 Homogeneity of Variance and the F-test 132
7.5 Difference of Means (independent samples) 135
7.6 Summary 141

Chapter 8 CATEGORICAL DATA
8.1 Introduction 143
8.2 Goodness-of-Fit 144
8.3 Test of Independence 148
8.4 Summary 152

Chapter 9 CORRELATION ANALYSIS
9.1 Introduction 153
9.2 Correlation 154
9.3 Test of Significance of Pearson Coefficient 157
9.4 Regression 158
9.5 Variation 161
9.6 Summary 162

Appendices
Appendix 1 - Data Set: Vital Statistics of High School Soccer Team 164
Appendix 2 - Normal Probability Distribution 165
Appendix 3 - t Distribution 166
Appendix 4 - Chi-Square Distribution 167

Index 168

CHAPTER 1
INTRODUCTION

SECTION 1 -- INTRODUCTION

What is statistics? The term statistics means many different things to different people. However, most of the definitions would fall into three general categories.

In the first general category, statistics can be thought of as numerically oriented facts. Examples of this type of statistics include the following statements: *The airline distance between New York and Auckland, New Zealand is 10,194 statue miles* and *In the 1888 presidential election, Benjamin Harrison won 233 electoral votes with 47.9% of the popular vote. In that election, Grover Cleveland, the incumbent, lost the presidency with 168 electoral votes, but 48.6% of the popular vote.* These statistics, and other similar statistics, are important. However, one has a sense that there is more to it than just this.

Answers in the second general category imply that statistics is a way or technique used to win an argument when reasoning and logic fail. Statisticians take a few numbers, put them into a little black box, turn the crank or press a button, and come up with a magical answer. Without an understanding of what is going on in that little black box, many people have the intuitive belief that the statistician is trying to put something over on them. People cannot trust something they cannot understand. Some people place more trust in alchemy than in statistics. Former British Prime Minister Benjamin Disraeli is reported to have said, "There are three kinds of lies: lies, damned lies and statistics." Unfortunately, there are enough examples of statistics being used in this manner, either intentionally or unintentionally, to give validity to this belief. The way to eliminate this fear of statistics is through education.

The following letter to Ann Landers, which appeared in the **Sunday Record**, of Middletown, New York, on June 24, 1990, illustrates how some people view statistics.

GOOD STATISTICS?

Dear Ann Landers:

If you have any clout with scientists, will you please tell them to lay off? Up 'til now they've been fairly sensible, but there's such a thing as going too far and I think they have done it.

I read about some double-dome experts who have discovered that when the calorie intake of rats is sharply reduced, they live a great deal longer. Translated into human terms, the scientists say, if people did this they might expand their life expectancy to 170 years.

Sound wonderful? Well, think it through and you will conclude that it would be a nightmare without end. For example: For Christmas dinner, you would have to include not only your in-laws, but a couple sets of grandparents, great-grandparents, great--great-grandparents and great-great--great-grandparents. Your dining room couldn't possibly accommodate that crowd, so you would have to rent a hall.

If you were 170 years old, you would have to buy gifts for your grandchildren, great-grandchildren, great-great-grandchildren and great-great-great-grandchildren. This would be quite a financial burden, especially for a person who hasn't been employed for 100 years. And where, please tell me, would these 120-,130-,140-, 150-, 160- and 170-year-old people live? Today it's not easy to find a good nursing home for less than $30,000 a year. And don't forget that you'd have to take care of not only your aged parents but their parents, their parents' parents, and their parents' parents' parents. Sound frightening? Wait, there's more.

How would you like to pay alimony for 100 years or more? If science doesn't stop coming up with "remarkable discoveries," it just might happen.

And what about Social Security? Forget it. There wouldn't be any. No system could accommodate the financial burden.

While the laboratory rats were in pretty good physical condition when they reached what in rat terms was 170 years, what about their mental faculties? How do you give a rat an IQ test? Most people become forgetful in their late 60s. What would life be like with millions of people loose for an added 100 years who can't read a road map or remember their own telephone number? And the post offices would be in for major bedlam. Can you imagine what the handwriting of these 170-year-olds would look like?

How would you like to see politicians running for a 15th term? With people living beyond 150 years of age, you can be sure that those who are interested in elective office would hang around as long as they had a pulse.

I am a 60-year-old woman who had a face lift 10 years ago and I still look pretty good. Ann I don't want to think what I will look like if I should live another 100 years. I enjoy life, but enough is enough.

Realist in Virginia

[used with permission of Ann Landers, Chicago Tribune]

Did you notice that there was a grammatical shift between the first and second definitions? In the first definition, *statistics* is plural. **Statistics are numerically oriented facts.** In the second definition, *statistics* is singular. **Statistics is a methodology.**

The third category carries the idea of a methodology even further, by formalizing the process and attempting to place a firm foundation under statistics. In this category, we look at how data is gathered, how it is reported, and what the data means. We define statistics as follows:

DEFINITION 1.1.1
STATISTICS is the process of collecting, presenting, analyzing and interpreting data.

Within this definition, there are three different aspects of the process. These aspects have evolved their own separate identities. Differing techniques have been developed in each of these areas.

1. **RESEARCH DESIGN** involves the collection of data. "How do we find or gather the data with which we work?"

2. **DESCRIPTIVE STATISTICS** involves the organization and presentation of data. "How do we organize and present our data in a way that makes sense?"

3. **INFERENTIAL STATISTICS** involves the analysis and interpretation of data. "What does our data mean?"

Statistics may be viewed as a tool for other disciplines by which an understanding of our world may be sought. Examples and problems throughout the text will use applications of statistics in business, economics and management sciences; the social and behavioral sciences of psychology, sociology and history; and many others. These examples and fields are not exhaustive. The methods and techniques discussed will have broad applications to many areas. We hope that this text will serve as a road map to you indicating directions in which you may proceed in your own chosen fields with your own problems.

HISTORICAL NOTE

The use of statistics, numerically oriented facts, is as old as writing, if not older. Shepherds of the ancient world, who could not count as we know counting, knew if even one sheep were lost. Sailors in the ancient Mediterranean kept track of their cargoes. Money lenders in ancient Egypt kept track of their accounts. In the Old Testament book of Genesis, there are references to the ages of individuals, or to the population of cities, or to the accounting of the wealth of individuals. These are all uses of statistics.

Many statisticians think modern statistics began in 1662 with the publication of "Natural and Political Observations Made upon the Bills of Mortality." John Graunt (1620-1674) was a successful businessman in seventeenth century London. His business was so successful that he had "time on his hands," and he could pursue other interests. One of these outside interests was the reading and studying of the "Bills of Mortality," a weekly church listing of births, christening, and deaths and their causes.

Graunt was interested in the difference between the birth and death rates of men and women. His study of the massive set of raw data led him to conclude that there was indeed some consistency among the birth and death rates of each sex. Many statisticians recognize this study as the first published interpretation of a real phenomenon based upon on an analysis of data. This was the birth of statistics.

Graunt's claim is not without detractors. Tycho Brahe (1546-1601) spent years collecting a vast array of immensely accurate, astronomical data on the positions of the planets. With his sudden death in 1601, this data was inherited by his assistant Johann Kepler (1571-1630). Kepler spent eighteen years analyzing this data, before publishing in 1619, his monumental work Harmony of the Worlds. In this work, Kepler outlined three laws of planetary motion, which would revolutionize astronomy, physics and even mathematics. Kepler's work of analyzing Brahe's data and empirically formulating an explanation of that data was one of the most extraordinary events in science. What did it involve? The collection and organization of data (Brahe), and the presentation, analysis and interpretation of the data (Kepler).

SECTION 2 -- DEFINITIONS

In the study of any area, to provide a consistent basis of discussion, we must agree upon the definitions of certain terms used repeatedly. Our study of statistics is no different. To introduce some terms that we will use in this text, consider the following example:

EXAMPLE 1.2.1: An industrial firm is engaged in the production of light bulbs. The firm is required by law to state on any packaging the brightness or intensity of the bulbs and the life expectancy of the bulbs. If the firm is of any size, it will have more than one machine manufacturing light bulbs. Due to slight differences in machines, it is reasonable to expect slight differences in the product of different machines. Thus, it is reasonable to expect that different light bulbs will shine with different intensities and last different lengths of time. [QUESTION: Even if there were only one machine, would all the light bulbs be the same?] Intensity and life expectancy vary from light bulb to light bulb. Intensity and life expectancy are said to be VARIABLES.

DEFINITION 1.2.1
A **VARIABLE** is a characteristic or phenomenon which may take on different values and which may be measured or counted in some manner.

To determine the life expectancy of their light bulbs, the firm could test every light bulb to see how long it would take until they burnt out. The firm would use up or exhaust the entire POPULATION.

DEFINITION 1.2.2
A **POPULATION** is a complete set of individuals, objects or measurements which have some common observable characteristic.

Would burning every light bulb out be economically feasible to the firm? Obviously, a firm approaching a problem in this manner would soon go bankrupt. A more sensible approach to the problem would be to select some bulbs from the daily output of light bulbs and carefully measure their lighting life. These selected bulbs are said to form a SAMPLE.

DEFINITION 1.2.3
A **SAMPLE** is a subset of a specified population

The process of selecting a sample and measuring the life span of each bulb in the sample is called an EXPERIMENT.

DEFINITION 1.2.4
An **EXPERIMENT** is the process of selecting a sample and measuring some variable of the sample.

The measurement or observation of the time required for the sample light bulbs to burn out generate a collection of times. These times or numbers are said to be DATA.

DEFINITION 1.2.5
DATA are numbers or measurements which are collected as a result of observations.

The data on life expectancy may be manipulated according to certain rules to yield summary statements, such as the "average" life span of the sample light bulbs. The resulting numerical value is a STATISTIC.

DEFINITION 1.2.6
A **STATISTIC** is a number resulting from the processing of data according to specified techniques or strategies, which describes a characteristic of the sample.

In this example, the average life span of a sample of light bulbs is a statistic. Does this statistic suggest anything about the life expectancy of all light bulbs in the entire population? Intuitively, certain samples and the corresponding statistic of average life span will say more about the population life expectancy than with other samples. [QUESTION: Can you describe why some samples would say more about the population than others?] This population life expectancy is called a PARAMETER.

DEFINITION 1.2.7

A **PARAMETER** is any characteristic of a population which can be measured or observed, and described quantitatively or numerically.

Although it is physically POSSIBLE to determine the parameter in this example, it is neither economically nor practically feasible to do so. Parameters are RARELY known. Parameters are, however, commonly estimated from sample statistics. This process is essentially the heart of INFERENTIAL STATISTICS. Within this example the three areas or functions of statistics are apparent.

DEFINITION 1.2.8

RESEARCH DESIGN is the planning and implementation of an experiment, and the gathering of appropriate data.

DEFINITION 1.2.9

DESCRIPTIVE STATISTICS is the presentation of information, derived from data, in a convenient, usable and understandable format.

DEFINITION 1.2.10

INFERENTIAL STATISTICS is the generalization of information derived in the descriptive process; for example, making statements about the whole knowing only the part.

The entire statistical process may be diagrammed as follow

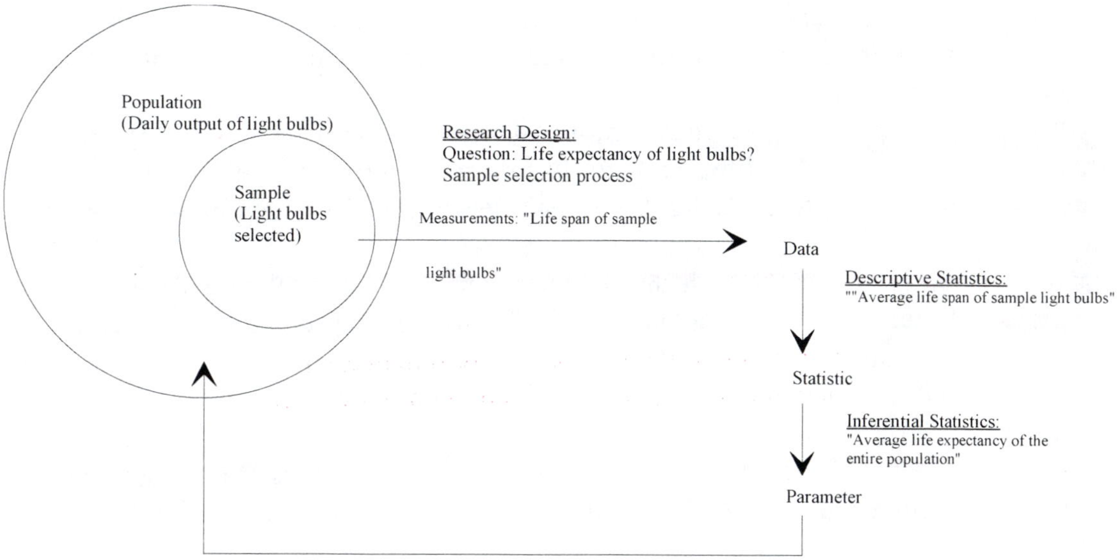

DIAGRAM 1.2.1: RESEARCH DESIGN is used to select one or more SAMPLES from a POPULATION, to investigate one or more VARIABLES. DATA are collected and summarized using DESCRIPTIVE STATISTICS. INFERENTIAL STATISTICS is used to estimate one or more population PARAMETERS from the sample STATISTICS.

SECTION 3 -- RESEARCH DESIGN

Every problem or process must have a starting point. If statistics is the process of collecting, organizing, presenting, analyzing and interpreting data, then statistics must begin with data. There are two fundamental questions that must be considered: 1) What is data? 2) How is it obtained?

Data is a peculiar word. It is technically a plural noun. The word "data" has been lifted directly from the Latin. It is the plural form of the word "datum," which meant "fact." In English usage today, most people ignore the singular datum, and refer to everything as data.

The definition given above for data says that data are numerically oriented facts. These facts typically are counts or measurements, depending upon the problem under consideration. Examples would include the number of two-family houses on a given street, the order of finish in a race, the temperature of a liquid mixture, or the weight of an individual. Each problem will determine the type of data to be considered, and intuitively, give some insight in how to obtain this data.

Therefore, if the purpose of research design is the collecting of data, research design must begin with the problem. Consider the problem of determining the number of students enrolled at Ivy Hall University. You might be tempted to say, "That's easy to solve! You just count them."

However, stop and think for a moment. What does it mean to be enrolled? Does it mean "taking classes?" Does it mean "to be registered in the Registrar's Office?" Does it mean to have completed the official admission process and to have been formally accepted into the University? The answers to these three questions might be different, and hence the number of enrolled students would vary according to interpretation.

The most important aspect in research design is the definition of the problem. "What is the problem? What do I want to know?" If the problem has been properly defined, then usually you know what data needs to be collected. At least, you can determine some idea of what data you should collect.

The unfortunate thing is that for each different type of data, there is generally a different method of gathering it. What we have attempted to do in this text, is to present some standardized or typical problems, show you how to deal with these situations, and indicate how you might translate these methods to the particular disciplines or problems with which you are working. This forms the material of Chapter 2.

The importance of design can never be overemphasized. **Design is the thing.** This is illustrated very well by Light, Singer and Willett, in the preface to their book, *By Design: Planning Research on Higher Education*,

"We emphasize research design over measurement and analysis. This is because good design comes first. No matter how precise your measurement or how sophisticated your analyses, you risk failure if your research is not well planned. You can't fix by analysis what you bungled by design." [pages vii and viii.]

SECTION 4 -- DESCRIPTIVE STATISTICS

Descriptive statistics can be intuitively defined as the process of organizing and presenting data in useful forms or formats. Although, as we will see, there are many different ways of organizing data, there are typically two general means of presenting it. They are numerically, as in a table, or visually, as in a graph. The type of data and how it is gathered will affect how we organize it. In working with data, there is not always a right and wrong way to organize it. Some ways may be more effective than others, given a specific goal you have in mind.

Similarly, in the presentation of data, different problems and different goals may suggest different ways of presenting the same data. The people who will read and use your data can also affect the manner in which the data is presented. Some people want to see the information in a graph or chart, while others want to see the "real numbers" in a table. You need to know not only your data, but also your intended audience.

As you can see, there is no one way to organize and present data. What we attempt to do in this text is present some standard ways of organizing and presenting data. It is our hope that you will be able to apply these ways to the problems with which you will deal, or that you will be able to modify these, or even generate new ways to apply to your data. A significant section of the text is devoted to this area called descriptive statistics. The following example may help to explain what you will be doing in this section:

EXAMPLE 1.4.1: In selecting applicants for admissions to the highly competitive School of Law at Ivy Hall University, the Dean of Admissions has several measuring instruments from which to choose. These instruments might include the student's undergraduate grade point average and rank in class, the student's score on the Graduate Record Examination (GRE), the student's score on the Law School Admission Test (LSAT), and the student's score on the Law School Data Assembly Service (LSDAS). How can the Dean use the information available to make the best decisions for the Law School and for the students involved?

1. One way the Dean could use the scores would be to rearrange them from the highest to lowest, and then group them into useful categories to obtain an "overall picture" of the data at a glance. From this grouping, the Dean could construct tables, graphs and figures to allow easier visualization of the results. (Chapter 3: Data Presentation)

2. The Dean could calculate averages to learn something about the "typical" performances of the students or the performance of "typical" students. (Chapter 4, Section 2: Measures of Central Tendencies)

3. Employing the average as a reference point, the Dean could describe the dispersion of scores about this central point to get an idea of the spread of the scores. (Chapter 4, Section 3: Measures of Dispersion)

4. In order to compare the GRE and LSAT for individual students, the Dean could convert the raw scores of each test into percentiles. (Chapter 4, Section 4: Measures of Location)

5. The Dean might want to place an individual's score into the context of the entire distribution in order to see how that individual compares with the group. (Chapter 5: Probability and Sampling Distributions)

6. The Dean might want to try to determine if there is some relationship between a student's entry data and that student's grade point average while in law school. (Chapter 9: Correlation Analysis)

In all these cases the key is to ask the right questions. "What do I know? What do I want to know? What is the easiest way to get from what I know to what I want to know?" It is our intent to help you learn the right questions and some of the right answers.

SECTION 5 -- INFERENTIAL STATISTICS

Inferential statistics can be intuitively defined as the process of analyzing and interpreting data to make statements about the unknown from the known. Much of the material discussed in descriptive statistics really forms a bridge to the inferential. The final portion, Chapters 5 through 9 of this text, is concerned primarily with the formation and testing of hypotheses, or educated guesses, about data. How do we know what guesses to make? Depending upon the data, there are many different directions that could be taken. However, no matter what direction we take, there are several unifying themes. To illustrate this point, consider the following example:

EXAMPLE 1.5.1: A pharmacologist is interested in determining the effects of a vaccine on the occurrence rate of Hong Kong flu among a given population. The pharmacologist sets up two vaccines, one real and one a placebo, and inoculates half of the population with each vaccine. After two months, the "real vaccine group" has a lower incidence rate of Hong Kong flu. Did the vaccine work? It is tempting to say, "Yes, the incidence rate was lower." The incidence rate might well be lower, but that does not answer the question, "Did the vaccine work?" Even if the drug had absolutely no effect, it is highly unlikely that the incidence rates of the two groups would be identical. Some differences would occur naturally. The critical questions are:

1. Is the difference between the two incidence rates large enough to rule out random variations?

2. Does the pharmacologist have enough confidence and evidence to predict the outcome of a repeat of the experiment with a different population? How much confidence?

In working with inferential statistics, we do not wish to sound like a broken record, but the key is to ask the right questions. "What do I know? What do I want to know? What is the easiest way to get from what I know to what I want to know?" If you can ask the right questions, the right answers will come.

SECTION 6 -- ETHICAL CONSIDERATIONS

The second category of answers to the question "What is statistics?" involves the idea that statistics was something less than ethical. In the first place, statistics can neither be ethical nor unethical. People are ethical or unethical. It is not our intention to teach you how to be unethical, but to help you spot what might be inappropriate uses of statistics by others and to avoid your own inappropriate uses of statistics. Sometimes these inappropriate uses are obvious.

EXAMPLE 1.6.1: A magazine advertisement claims "8 out of 9 podiatrists surveyed recommended BOILS' Footpads." How many doctors were included in the survey? The ad does not really say. If 1000 podiatrists were surveyed and 889 recommended BOILS', that is impressive. However, it is possible that only 9 doctors were surveyed, and 8 of those worked for BOILS'. It would be very difficult for them not to recommend the company's footpads. Even if a large number of doctors were included in the survey, the survey could have been biased by the way the questions were asked. The question "What footpad do you recommend?" is quite different from "Would you recommend BOILS'?" This ad leaves much to the imagination.

Sometimes the questions are much more subtle. Consider the Associated Press newspaper article from the Lebanon Daily News of July 31, 1985 that is reprinted below.

Pa. Examples Cited
FBI's Crime Statistics Misleading

STATE COLLEGE, Pa (AP) - Recently released FBI statistics showing a higher per capita crime rate in the State College area than in Pittsburgh do not necessarily mean the Centre County region has more crime than the big city, officials said.

The FBI on Monday released its national survey of serious crimes per 100,000 population for various metropolitan areas in each state last year.

The crime index said the eight-county Philadelphia area had 4,162 serious crimes per 100,000 population in 1984.

The State College area, including Centre County, followed with 3,383. The four-county Pittsburgh area reported 3,243.

Several factors could have distorted the State College figures, said Ned Witmer, supervisor of the Uniform Crime Reporting Section for the state police in Harrisburg.

In calculating the index, the FBI did not include the more than 33,000 students who attend Pennsylvania State University in State College, he said.

Witmer said the FBI index does not show that violent crimes--murder, rape, robbery and aggravated assault--accounted for only 2.7 percent of the State College area's total crimes, against 10.2 percent in the Pittsburgh area.

Better reporting of crime could also account for the higher-than-expected crime rate in Centre County, Witmer said.

"In suburban areas more so than urban, citizens are very adamant about what's theirs is theirs" and are more likely to report minor thefts, he said.

"You can get numbers that say anything you want," Witmer said of the index.
For example, rural Forest County with an approximate population of 5,000, "perpetually has the highest crime rate of any area in the state," he said.

Its 1984 crime index per 100,000 was

5,832. In actuality, Forest County had only 298 serious crimes, most of which were burglaries in unoccupied hunting cabins

"So using the per capita, there's a lot of crime in Forest County," he said.

The statistics included in this article show a higher per capita crime rate in the State College area than in Pittsburgh. The lead paragraph asks how the Centre County region can have more crime than the big city.

The article offers several explanations for this apparent discrepancy. First in calculating the index, the FBI did not include the 33,000 students of Pennsylvania State University. This is a significant portion of the population of the area. Although the overwhelming majority of college students are not involved in criminal activities, a certain number are arrested each year.

The second explanation centers around the hypothesis that there is better reporting of crime in the suburban and rural areas than in the urban areas. The article cites rural Forest County with a population of 5000. It "perpetually has the highest crime rate of any area in the state." The 1984 crime index of 5,832 crimes per 100,000 people was determined by 298 actual serious crimes, most of which were burglaries in unoccupied hunting cabins.

Was the misuse of statistics in EXAMPLE 1.6.1 intentional? It is impossible to decide that. However, the ad is so questionable that anyone reading it should beware. We would hope that no one would knowingly misuse statistics in such a way. Was the misuse of statistics in the newspaper article intentional? We do not believe it was. However, it does illustrate how easily statistics can be misused. What we would hope is that after studying statistics, you would be able to recognize the problems and shortcomings of your own statistics, and the statistics of others.

The following Associated Press article that appeared in **The Enterprise**, Brockton, Massachusetts, on May 26, 1985, illustrates just how far some people will go.

What is right and what is wrong?
Ethics of reporting results

Columbia University retracts 10 erroneous scientific studies

Boston (AP) - Ten scientific studies conducted at Columbia University are being retracted from medical journals because their results are wrong, and a committee investigating the affair said it could neither prove nor rule out "deliberate fabrication of data," officials say.

The committee, headed by a Nobel Prize winner, criticized a young researcher for continuing to rely on a test for four years after doubts were raised about its accuracy. And it chastised a senior physician for listing himself as a co-author of studies that he was not directly involved in.

The controversy centers on Dr. S. Alexander Stalcup, acting head of the vasoactive mediator research lab at Columbia's College of Physicians and Surgeons, and Dr. Robert B. Mellins, chief of the pediatric pulmonary division.

The problems came to light after the first three reactions were printed in the May issue of the Journal of Clinical Investigation. The Boston-based journal published the discredited studies between 1979 and 1984.

Dr. Henrik H. Bendixen, dean of the medical school, attributed the problems in part to the "high-pressure environment" at Columbia.

Stalcup blamed the discrepancies on a difficult test, and he denied any fraud was involved.

According to the committee report, Stalcup, 39, brought the test to the lab in 1975. He used it to develop a theory about how the lungs produce chemicals that regulate blood pressure, especially when they are deprived of oxygen. The research had implications for the treatment of heart attacks and lung disease.

Research by scientists at other universities conflicted with Stalcup's. And in 1980, the committee said the test stopped working consistently. However, Stalcup and his boss, Mellins, continued to produce papers through 1983 that confirmed the theory and were based, in part, on the dubious test.

Last year, Stalcup developed a new, apparently more accurate test. It refuted the results of the earlier test. When this was learned, Columbia named two committees to investigate. An internal committee was comprised of faculty members and a five-member outside committee was headed by Dr. Rosalyn Yalow of the Bronx Veterans Hospital, who won the Nobel Prize in medicine in 1977.

"The external review committee could not find overt evidence for deliberate fabrication of data, although this could not be excluded," the committee wrote. "However, the committee established that Dr. Stalcup had not adequately supervised the technical staff.

"In addition, the continued publication and presentation of papers based on data of dubious certainty showed poor professional judgment. The committee found Dr. Stalcup's written and oral explanation of his actions contradictory and confusing."

The committee noted that since 1978, about half of Mellins' published papers involved the disputed work.

"Despite this deep involvement, Dr. Mellins appeared to be unaware of the problems underlying the papers he co-authored," the committee said. "Although it is not uncommon for senior scientists to co-author papers without direct participation in the research, this practice is not in the best interest of science."

The committee also criticized Mellins for not supervising Stalcup more closely.

Stalcup eventually developed another test using tritium, a more stable radioactive material, and repeated the studies a third time. Stalcup, an associate professor, said he is looking for a new job for personal reasons, and Bendixen, the dean, said he has not been disciplined.

"We are living in a high-pressured environment, and that was a contributing factor," Bendixen said. "The modern universities, particularly the more prestigious ones, are pressure cookers."

Besides the Journal of Clinical Investigation, the university has asked that studies be retracted in Circulation, Circu-lation Research and the Journal of Applied Physiology.

Among those whose work was retracted is Dr. Hugh M. O'Brodovich, now of McMaster University Medical Center in Hamilton, Ontario.

I spent three years and worked very hard," he said. "All that work's down the tube now."

Stalcup was awarded a five-year, $180,000 grant from the American Heart Association that was scheduled to be finished June 30. Howard L. Lewis, an association spokesman, said the final, $7,500 payment has been withheld because of the controversy.

Asked if he had done anything wrong, Stalcup replied, "I wish we had developed newer technologies earlier.

"We all feel devastated by it," he added. "This has been a painful episode."

SECTION 7 -- SUMMARY

In this chapter some key terms and ideas regarding statistics were introduced. In particular, three functions within statistics were identified, and associated techniques within each function were introduced. In addition, an outline of the text was presented. Ethical concerns were also raised.

KEY TERMS TO KNOW

Statistics
Research design
Descriptive statistics
Inferential statistics
Variable
Population

Sample
Experiment
Data
Statistic
Parameter

CHAPTER 2
RESEARCH DESIGN AND DATA COLLECTION

SECTION 1 – INTRODUCTION

In Chapter 1, we indicated that statistics is the process of collecting, organizing, presenting, analyzing and interpreting data. This process must begin with data. There are two obvious questions. 1) What is data? 2) How do we obtain data?

In Chapter 1, the first question was answered simply by saying that data are numerically oriented facts. Intuitively most people will have some understanding of what is meant by this definition. However, this answer also has a great deal of depth to it. What is a number? There are at least three types of numbers: nominal, cardinal and ordinal.

Nominal numbers, as the term suggests, name things. The symbol "8" on the back of the jersey of a soccer player is a nominal number or numeral. Nominal numbers may or may not have other meanings associated with them. The main purpose of nominal numbers is to identify objects.

Cardinal numbers are used to answer the question "How many?" In outdoor soccer, teams play with 11 players. Many soccer teams will have 18 or more players on their rosters, but only 11 of them play at any one time. The numbers "11" and "18" are examples of cardinal numbers. Cardinal numbers are used to count objects. In statistics, cardinal numbers are generally more useful than nominal numbers. Unfortunately, statistics in its early development did not always "follow" mathematics. Problems which involve the division of objects into distinct categories and the determining of the number of objects in each category have been called "nominal" problems instead of "cardinal" problems. There will be several times when one word will have two meanings. With practice, the student should be able to determine which meaning is intended from the context. This should not be that surprising, since many words have more than one meaning.

The name "ordinal" itself suggests the identifying characteristic of the last type of number. There is an underlying arrangement, or "ordering," associated with ordinal numbers. This ordering may be the natural ordering of the number line (1 is less than 2, 2 is less than 3, 4 is less than 4.69); it may be the natural derived ordering (first, second, third, etc.); or it may be another arrangement, such as military or professorial rank.

Data can also be broadly divided into two general types, discrete and continuous. The difference between these types can best be described by looking at the general means of data collection associated with each. Problems in which counting methods are used deal with discrete data. Problems in which measurements are taken are concerned with continuous data. That seems simple. However, sometimes it is not so easy to decide with which type of data you are working. On occasion, it is necessary to approximate one type of data with the other.

In this chapter, we will consider the two types of data. Section 2 will introduce discrete data and the two different levels of discrete data. Section 3 will introduce continuous data and two different levels of continuous data. In Section 4, data collection will be considered. Problems associated with large data sets will be discussed and possible solutions suggested. The idea of sampling will be treated more fully in Section 5.

SECTION 2 -- DISCRETE DATA

Discrete data is typically divided into two levels that are associated with two different types of numbers. The first type of data is concerned with the number of objects that are in distinct categories.

EXAMPLE 2.2.1: In Ivy Hall University, there are 10,661 students. These students may be classified according to gender. The following table gives such a classification:

GENDER	NO. OF STUDENTS
Male	4,427
Female	6,234
Total	**10,661**

EXAMPLE 2.2.2: These same students might also be classified according to the federal breakdown of ethnic origins. The following table gives such a classification for Ivy Hall University:

ETHNIC GROUP	NO. OF STUDENTS
International	131
Black Non-Hispanic	1,234
American Indian/Alaskan	23
Asian or Pacific Islander	132
Hispanic	820
White Non-Hispanic	8,306
Other	15
Total	**10,661**

This type of classification of information into categories is called categorical or **NOMINAL** measurement. Nominal data consists of the count of individuals or items in each of two or more categories. The term *nominal* comes from the Latin word for *name.*

In the two examples above, there are three properties of the categories that should be apparent. The first property is that the categories share no individuals. A person cannot both be a male or a female at the same time. For ethnic origins, the federal regulations require colleges and universities to count a student in only one category. When the characteristic of interest can be counted in only one group or category, the categories are said to be **MUTUALLY EXCLUSIVE**.

The second property of the categories above is that there is no natural ordering of the categories. One gender is not better or above the other gender. One ethnic group does not come before the other groups. The third property of the categories above is that every student is accounted for in the classification process. Each student must be a male or a female. Each student must have an ethnic group. If it is not one of the six that the federal government recognizes, the student is classified as "Other." When each individual must appear in one of the categories, the categories are said to be **EXHAUSTIVE**. The above suggests the following definition:

DEFINITION 2.2.1

NOMINAL DATA are data which are "measured" by dividing them up into mutually exclusive, exhaustive categories, which have no natural ordering.

What happens when there is a natural ordering which may be imposed on the categories? Consider the example of Ivy Hall University again. This time the students are divided into classes. The breakdown is given the following table:

EXAMPLE 2.2.3:

CLASS	NO. OF STUDENTS
Freshman	2,951
Sophomore	2,267
Junior	1,931
Senior	1,732
Graduate Students	1,780
Total	**10,661**

The categories above are mutually exclusive and exhaustive. A student cannot be in two different classes simultaneously. Each student must be classified. The difference between this classification and the gender classification is that there is a natural ordering associated with the categories. Sophomores are "above" freshmen. Graduate students are "above" seniors. This example suggests the following definition:

DEFINITION 2.2.2
CATEGORICAL ORDINAL DATA are data which are "measured" by dividing them into mutually exclusive, exhaustive categories, which have a naturally associated order.

There is a second type of ordering in reference to discrete data. An example of this type of ordering is the "order of finish" in a race. What are the characteristics of this type of ordering? Each individual or object has one and only one place. In a swimming race, each swimmer is assigned a lane. A swimmer cannot swim out of his or her lane, under penalty of disqualification. If a swimmer gets a cramp and stops swimming, that lane is still assigned and cannot be used by another swimmer. In a sense this ordering can be said to be mutually exclusive and exhaustive.

You may be asking, "What about ties?" In a race, it is possible for two to finish at "the same time." There are several different ways to handle ties, and it usually depends upon the context of the data, how they will be handled. These methods will be discussed as they arise in the text. With the understanding that ties must be considered, it is possible to make the following definition:

DEFINITION 2.2.3
INDIVIDUAL ORDINAL DATA are data which are "measured by dividing them into mutually exclusive places, which have an associated "order."

SECTION 3 -- CONTINUOUS DATA

In mathematics, a distinction is made between the discrete and the continuous. A discrete variable can only take on individual values that are distinct, disjoint, and separate, such as the number of vehicles that go through an intersection in any given hour. A continuous variable can take on values that "run together," such as the straight-line equation $y = 2x + 1$, defined on the real line. Sometimes the distinction between discrete and continuous is blurred. Is the weight of an object a discrete variable or a continuous variable? Most people talk about their weight in terms of pounds, which are distinct, disjoint and separate numbers. However, with a very accurate scale, the weight of an individual could be measured to tenths or even possibly hundredths of a pound. It would be as foolish to talk about measuring an ocean vessel to that accuracy as it would be to consider weighing silicon computer chips to that accuracy. Weight is a characteristic which can be measured to as accurate a degree as "is desired." Theoretically, weight can be measured to an infinite degree of accuracy; that is, given an accuracy of weight, one more degree of accuracy is possible. The example of a person's weight above illustrates this point. Given a weight in pounds, it is theoretically possible to weigh the person to the nearest tenth of a pound. This fuzziness between the discrete and continuous will enable the statistician on occasion to move back and forth between discrete and continuous methods.

There are two types of continuous data. The major distinction between the two types is not dependent upon ordering. The difference is much more subtle. Consider the following two examples:

EXAMPLE 2.3.1: The temperature of a solid metal rod.

EXAMPLE 2.3.2: The time it takes to run 100 meters.

What is the difference between these two examples? Why is time so different from temperature? The easiest way to explain the distinction is to look at some possible values for these two examples. In EXAMPLE 2.3.2, a world class runner would be running 100 meters in approximately 10 seconds. An elementary school child may run it in 20 seconds. We say that the athlete is twice as fast as the child is. In EXAMPLE 2.3.1, a solid metal rod placed in an oven may reach a temperature of 200 degrees, while that same rod placed in the sun's rays may reach a temperature of 100 degrees. Normally, the rod in the oven is not called twice as hot as the rod in the sun. Why can the word "twice" be applied to one example but not the other?

Consider the temperature example again. A temperature of 200 degrees would be measured on what is known as the Fahrenheit scale. This is not the only scale against which temperature can be measured. A temperature of 200 degrees Fahrenheit is the same as a temperature of 93.3 Celsius, while a temperature of 100 degrees Fahrenheit is 37.8 degrees Celsius. On the Celsius scale, 93.3 is not twice 37.8. Is this the only difference?

Consider the time example. Is there an absolute minimum time possible? Although it appears that the human body can be trained in a way that keeps lowering the time it takes to run 100 meters, no one will **EVER** run 100 meters in less than 0 seconds. There is an absolute zero in working with measured time. The number "0" in the time example is fixed. In the temperature example, depending upon the scale used, "0" is arbitrarily placed. These two distinctions form the basis for defining the following two categories of continuous data:

DEFINITION 2.3.1

INTERVAL DATA are continuous data which are measured against a scale which has an arbitrarily placed zero. **INTERVAL DATA** also have the property that a unit difference is constant throughout the range of possible values.

DEFINITION 2.3.2
RATIO DATA are continuous data which are measured against a scale which has a fixed zero. **RATIO DATA** also have the property that a difference ratio remains constant throughout the range of possible values.

Among the four types of measurements, there is an inherent ordering. There is a natural break between the discrete and continuous data measurements. **NOMINAL** data, measuring discrete data by just dividing it up into categories is considered the lowest level of measurement. **ORDINAL** data, which has an associated order, is one step up the measurement ladder. **INTERVAL** data, which is continuous data with an arbitrary zero, is next. **RATIO** data, with a fixed zero and data for which ratios make sense, is the highest level of data. This can be illustrated in the following diagram:

DIAGRAM 2.3.1

RATIO
$\Uparrow$
INTERVAL
$\Uparrow$
ORDINAL
$\Uparrow$
NOMINAL

Since we have different levels of measurement to use with data, this might suggest that there are different levels of statistical processes to use in handling this data. For the most part, this is true. The following convention is a good rule-of-thumb to use:

CONVENTION 2.3.1
Statistical processes and statistics based on one level of measurement may be used (if it makes sense) on a higher level of measurement, but should not be used (even if it appears to make sense) on a lower level.

Using this principle, we can, for example, calculate "averages" for interval or ratio data, but we should not calculate them for nominal or ordinal data. Reconsider the data in EXAMPLE 2.2.2. This data was the ethnic breakdown of the students of Ivy Hall University.

EXAMPLE 2.3.3: The following are the numerical labels suggested for the federal breakdown of ethnic origins:

ETHNIC GROUPS	LABEL
International	1
Black Non-Hispanic	2
American Indian/Alaskan	3
Asian/Pacific Islander	4
Hispanic	5
White Non-Hispanic	6
Other	7

Calculation of a "weighted average" using the numerical labels would be at best nonsense, and at worst, something else. In spite of the dangers and many warnings, this guideline has been ignored often, even by competent statisticians. One of the most common examples of where this guideline has been ignored is in the analysis of Likert scales. Likert scales are usually four or five point ordinal scales, such as the one illustrated below:

DIAGRAM 2.3.2

Superior	Good	Average	Poor	Very Poor
1	2	3	4	5

When numerical labels are placed on the categories, many people will treat the results as interval or ratio data, instead of the ordinal data that it is. Only in more advanced statistical methods is any distinction made in the treatment of interval and ratio data. Usually it is not necessary to identify whether the data under question has an absolute zero or not.

SECTION 4 -- DATA COLLECTION AND RESEARCH DESIGN

The process of data collection can be simply defined as the gathering of the information wanted or needed for the project at hand. Unfortunately, nothing is ever as simple as that. To gather the data desired, you need to know what data to gather. That is the process of **RESEARCH DESIGN**. In section 1.2, we defined research design as the "planning and implementation of an experiment, and the gathering of data from that experiment." Another way of stating the same thing is:

DEFINITION 2.4.1
RESEARCH DESIGN is the process of defining the problem under consideration and determining the optimum strategy for solving or answering the problem.

Typically, the statistician has the least input in this part of the statistical process. It is not reasonable to expect a statistician to be an expert in every discipline. The statistician cannot know management science, political science, sociology, psychology, biology, agricultural science, medicine, just to name a few of the possible areas in which problems could arise. However, the management scientist, the political scientist, the sociologist, the psychologist, the life scientist should know enough statistics either to solve the less complicated problems or to be able to talk to a statistician who knows enough of the discipline under consideration to intellectually discuss the problem.

The normal process of defining the problem involves several steps. The most important step in the process, however, is determining what information is needed or what questions must be answered. Once the questions are determined, often the remainder of the process is straight forward. Although it is our intent in this text to help you see what questions to ask, nevertheless, we are not experts in all the various disciplines that use statistics. (We are PROBABLY NOT experts in any of them.) Much of the work in this area must be done in the discipline itself. This will come with experience in the area, from studying typical problems in the area and from involving yourself in the discipline.

After the question is posed and the information needed is determined, in an industrial setting the next step would be to develop and specify alternate procedures for getting the information. This "industrial procedure" should be the norm everywhere. For most problems, there are different feasible routes to the answer. The task of the statistician or the person working with statistics is to find the BEST way to solve the problem. Each alternative must be weighed with advantages and disadvantages evaluated. In business and scientific applications one usually must base the derived strategy on both cost and importance of the information.

EXAMPLE 2.4.1: A firm that manufactures light bulbs must print on its labels the life expectancy of the light bulbs. How does the firm determine this value? It could "try" every light bulb it produces to see how long they last. Would this be a feasible solution? It would give the most accurate answer. However, the firm would not remain in business very long trying to sell "burned out" light bulbs. What should the company do? The company could select a group of representative light bulbs and test them to see how long it takes them to burn out. From the information provided by the group, or SAMPLE, judgments about all the light bulbs could be made. How accurate would these judgments be? It depends on how good the sample and generalizations are in representing the whole.

Cost and importance are difficult ideas to "get a handle on." As the complexity of the problem increases, the more slippery cost and importance become. For complex problems, planning is essential. Time spent in mapping out the strategy to attack a problem is never wasted. As suggested in EXAMPLE 2.4.1, sampling can be an important strategy in the solution of some problems. The next section deals with some sampling strategies.

After posing the question and settling on a methodology to answer the question, the next step in the process is to carry out your research design. This step is really not part of research design. However, there are two suggestions to remember in the carrying out of the design that are so important that we do not want to leave this section without mentioning them. The first suggestion should be considered common sense: **PAY ATTENTION TO DETAILS.** Many studies and research projects have been sabotaged by the lack of attention to details by the participants. A good research design can be ruined by sloppy data collection or analysis. The second suggestion could also be considered common sense: **PAY ATTENTION TO PEOPLE.** Many research designs require the cooperative participation of people. You must respect people and make sure that people can see and feel that respect.

Authors Light, Singer and Willett in their book ***By Design*** offer four principles in doing research. Their first principle is that "your study's design is the single most important factor that determines whether your findings will be scientifically first-class." When researchers seek assistance, they often have their data already in hand. They want to know what statistical analyses or techniques will give them the desired results. Usually an entire project would have yielded more useful and more convincing results if the design had been more carefully thought through before collection data began. You cannot fix a weak project design by using elaborate statistical analyses.

A corollary to this first principle is "that you should explore many design options before adopting any plan, especially a weak one." There will be times when practical constraints prevent you from carrying out the ideal design. You must resist the temptation to say: "If I can't do it right, it doesn't matter how I do it. I'll just do something quick and dirty." If your first-choice design is not feasible, look for a good second choice. Explore all possibilities, and then decide.

The third maxim is "pay enormous attention to detail" This may seem trivial, but it is only common sense. The little details will ultimately decide the credibility of your project. In the evaluation of a new process, for example, you may want to use a technique that uses randomization in the selection and assignment of subjects. But there are many different ways to randomize. You may have a choice between two different methods, each of which has the same *power*, but whose costs, in terms of time, effort and money, are quite different. "Paying attention to detail always has a high payoff."

Their final principle looks at the researcher's attitude as he or she approaches research. A great deal of research requires the willing participation of subjects. You should continually remind yourself, "that research must respect collaboration and cooperation." You need to remember the Golden Rule: "Treat others as you would have them treat you."

SECTION 5 -- SAMPLES AND SAMPLING

In the previous sections, two questions were posed in various forms. First, "What is a sample?" The definition said that a sample was any subset of a population. The second question was "Why use a sample?" The usual answer to this question is some form of the statement, "It is not feasible to use the whole population." But it should be obvious that some samples are better than other samples. What makes the difference?

There are two types of samples. These types are called PROBABILITY SAMPLES and NONPROBABILITY SAMPLES. The first example illustrates a probability sample.

EXAMPLE 2.5.1: In an urn there are 100 marbles of various colors. A person reaches in and without looking draws out 15 marbles. In this example, there is nothing that would prevent any one marble from being selected. This is a **PROBABILITY** or **PROBABILISTIC SAMPLE.**

The example above suggests the following definition:

DEFINITION 2.5.1
A **PROBABILITY** or **PROBABILISTIC SAMPLE** is a sample in which each element or object in the population has the opportunity to be included.

If in a PROBABILITY SAMPLE, each object in the population has the opportunity to be selected, the definition of a NONPROBABILITY SAMPLE is obvious.

DEFINITION 2.5.2
A **NONPROBABILITY** or **NONPROBABILISTIC SAMPLE** is a sample in which not every element or object has the opportunity to be included.

In a nonprobability sample, every element or object does not have the opportunity to be included. This may occur because inclusion in the sample is based on some desired property or solely on the judgment of the person selecting the sample. The following example illustrates this.

> **EXAMPLE 2.5.2:** A student hired to do market research in a shopping mall is told to approach males who appear at least 20 years old. From the population of all people, or even all males, this is a **NONPROBABILITY** or **NONPROBABILISTIC SAMPLE**. Not every male has the chance to be included. Those who are young teenagers would not be included. Sometimes there might even be a very young-looking individual in his twenties who would be bypassed.
>
> In doing the survey work, suppose this college student wanted to "get done quickly," so she invited some of her college classmates to come down to the mall to be interviewed. She "happens" to select these fellows. In this process the interviewer is introducing herself into the sample, and selecting the sample strictly on her own judgment. This again would be a nonprobabilistic sample.

The terms probabilistic and nonprobabilistic samples do not in and of themselves denote good and bad samples. The probabilistic sample in Example 2.5.1 happened to be "good." If the young lady in Example 2.5.2 did her job properly, the nonprobabilistic sample would be "good" for the purposes for which it was collected. If she took the shortcut of picking her friends, it would be a "bad" sample. What makes a sample "good" or "bad?" Before that question can be answered, the question "Why sample?" must be more fully investigated.

Returning to the light bulbs of Example 2.4.1, it was suggested that the manufacturing firm would have to sample because the test was consumptive; that is, the test used up the product. There are several tests of this general type. A taste test is one example. No one wants to "use" a food product after someone else has tasted it. Certain medical tests are of this type. An experiment that examines the embryonic development of chicken eggs is a consumptive test. To examine an egg, it must be opened, which stops the development of that egg.

Sometimes the experiment is not consumptive or destructive, but a sample must still be employed. Consider the problem of counting the number of hammerhead sharks in the Atlantic Ocean, or even an area of the Atlantic. Because sharks do not stay in one place very long, it is essentially a physical impossibility to count every one of them. Thus, a second reason for sampling might be the physical impossibility of checking every element of the population.

On occasion it might be physically possible to check the entire population, but the cost would be prohibitive. Consider the problem of deciding how many television sets were receiving a particular program on cable. The technology exists to install "counters" on every set tied to the cable system. However, the cost would be too great for the cable companies to bear themselves. The subscribers most likely would not want to pay for this, and many advertisers could not afford the increased costs of the advertising. The solution is to sample a selected group of subscribers.

Besides the physical and cost factors, time may be a problem in surveying the entire population. A politician who wants to know what the people in his district think about a particular subject cannot take the time to talk to every voter in the district. By the time that would be accomplished, the burning issue in the minds of the electorate would have changed.

The final reason samples are used may be the most important. Sample results are usually adequate for most purposes. It is obviously not worth the extra time, money and energy spent to talk to every voter in a district to learn that 73% of the voters do not favor the imposition of a sewer tax, when a sample of two thousand voters suggests that 68% are not in favor of the tax.

What about the times when sample results are different from the real results? Consider the problem of a poll taken before the election which shows one candidate winning, but in the election that candidate loses by a wide margin. The most famous occurrence of this event was the 1936 election and *The Literary Digest* poll. *The Literary Digest*, a popular magazine of the day, surveyed over 10,000,000 people, received responses from 2,376,523 people, and determined that Alfred E. Landon would defeat incumbent Franklin D. Roosevelt. However, Roosevelt won handily. The poll was off by almost 19 percentage points. What went wrong?

The size of the poll was not the problem. The timing of the poll was not the problem. [Question: Can you describe events that could turn an election around in a short time?] The problem was in the method of selecting who was to be in the sample. *The Literary Digest* mailed questionnaires to people who had registered automobiles or who subscribed to telephone service. Today, these lists would represent almost all socioeconomic classes. However, in 1936 that was not true. Telephone users and automobile owners were typically in the higher economic settings, and therefore, tended to be more "Republican" than the poorer people who did not have cars and phones. This latter group saw the "Democrats" as their ticket to prosperity. In addition, the survey required a postcard to be returned to *The Literary Digest*. The only people who normally return such surveys are those who are very committed to a particular candidate, and not the "middle of the road population." Thus, the sample group was not representative of the population of the entire electorate. It was a **BIASED** sample.

DEFINITION 2.5.3

A **BIASED SAMPLE** is a sample which is not representative of the population from which it was drawn.

A probabilistic sample is generally going to represent the population from which it was drawn, and therefore not biased. It will be said to be **UNBIASED.**

DEFINITION 2.5.4

An **UNBIASED SAMPLE** is a sample which is representative of the population from which it was drawn.

There are many ways to obtain an unbiased sample. The methodology of sampling tends to be dictated by the problem and the data itself. In any one given problem, not all these methods would be appropriate for that problem.

The first method of sampling is the most widely used and the simplest to understand and explain. Common examples of this sampling technique include drawing names from a hat or a drum, selecting marbles from an urn, or picking cases of light bulbs "at random" to test life expectancy. Such a sample is called a **SIMPLE RANDOM SAMPLE**.

DEFINITION 2.5.5.
A **SIMPLE RANDOM SAMPLE** is a sample in which each member of the population has exactly the same chance of being selected.

The major problem with a simple random sample is that typically, it is too cumbersome. It is extremely difficult to use a simple random sample to pick names "at random" from the telephone book to call to survey them concerning their political preferences. How can the methodology be "improved?"

We can improve the methodology by somehow systematizing the selection process. This suggests the following definition:

DEFINITION 2.5.6
A **SYSTEMATIC RANDOM SAMPLE** is a sample in which each member of the population is given an identifying label, such as a number. Then, an arbitrary starting point is chosen, together with a rule for selecting the next member of the sample.

For example, suppose the 27th entry on the second column of the left-hand page of the telephone book is chosen, and the same entry on subsequent pages are also selected. This would be a systematic random sample.

Instead of labeling every element in the population, suppose we divided the population into two or more groups or classes that share the same characteristics, such as gender or ethnic background, if we were talking about people. From each group we select a sample. The groups are called **STRATA** and the sampling technique is called **STRATIFIED RANDOM SAMPLING**.

DEFINITION 2.5.7
A STRATIFIED RANDOM SAMPLE is a sample in which the population is divided into subgroups or strata, according to some identifying feature. From each stratum, a sample is drawn.

The samples drawn in each subgroup could either be simple or systematic random samples. Often, the number selected from each stratum will be proportionate to the percentage population of the strata. For example, suppose a student opinion survey concerning library hours is conducted. If 35% of the students are freshmen, 25% sophomores, 20% juniors and 20% seniors, then for a sample of size 100, 35 freshmen, 25 sophomores, 20 juniors and 20 seniors should be included. The students in each class could be selected by a systematic approach, such as picking the 7th name on an alphabetic class list, and selecting every 20th name after that. The basic problem with stratified sampling is that "complete knowledge" of the population must be present to divide the population into strata. Without this knowledge we must use another type of sampling.

Another type of sampling is called **CLUSTER SAMPLING**. This type of sampling is more involved than the first three types. To explain cluster sampling, consider the United States census. In the census, is every individual in the United States physically counted? Obviously, the answer must be NO. The cost in terms of time, energy and money would be prohibitive, to say the least! What does the Census Bureau do? First, the Bureau has divided the United States into many, small geographical areas, which have common properties. These areas may be as small as a city block that contains one large apartment complex, to as large as a rural county that is primarily farm land. An attempt is made to approximate the entire population in the selection of geographic sectors, or clusters. [Why must this be an approximation? What could be done to make it a good approximation?] The Bureau then mails to every known residence in the United States a census survey. (The Postal Service helps with this job.) The Census Bureau does not expect to get all of the surveys back. [What are some reasons why surveys would not be returned?] The Bureau then selects a sample of areas from each of the geographic clusters, and sends census takers into these areas to take as accurate a physical count of residents as possible. [Why is this count not perfect?] The Bureau then uses the physical counts as a gauge to measure the probable error in mail return. [Is the error rate going to be the same for every cluster? Why?] This probable error is applied to each geographical area within a given cluster, to get an estimate of the population in that cluster. The population of the United States is then estimated from the population of the clusters. This suggests the following definition:

DEFINITION 2.5.8
In a **CLUSTER SAMPLE**, the population is divided into subsets or clusters. A sample of sectors or subareas from each cluster is selected. In each of the selected sectors, a census of all the members (or as many as feasible) of that sector is taken.

It is intuitively obvious that sampling error can be reduced by increasing the size of the sample [Why?] In cluster sampling, a very precise sampling of a subset of the population can be done. Often this process is less costly than taking a simple random sample with the same precision. With the census, the interviews and physical counts are concentrated in small geographical areas. This costs much less than having to send census takers all over the United States to interview a few families in each city and town. In fact, the savings due to cluster analysis is such that even when simple random sampling is possible, many researchers will still use a clustered sample. One note of concern is the choice of the clusters. Bias can be very easily introduced by the choice of clusters.

A fifth type of sampling uses results that are "at hand." This type of sampling is called **CONVENIENCE SAMPLING**. In this type of sampling, the researcher may use subjects in her class. For some questions, this may be satisfactory. For other questions, it may lead to very biased results. For example, if the researcher were investigating the trait of blue eyes, the sample of students in a class might be unbiased because having blue eyes has very little to do with being in a college or school class. [Unless you were in a very special class, such as a class in a school on an Indian reservation. Why would such a class be a "problem" for this example?] You would have a very biased sample if you were to ask college students what they thought about financial aid.

DEFINITION 2.5.9
A **CONVENIENCE SAMPLE** is a sample which draws upon results that are readily available (or convenient).

A type of specialized convenience sampling is called **HISTORICAL SAMPLING**. There are times when a researcher is called upon to analyze something that has already happened, and it is impossible or impractical to recreate it in a "laboratory setting." For example, suppose there has been an alarming increase in the incidence rate of lung cancer among male adults in a given population. How can a researcher find out why this has occurred? One method would be to look at a representative sample of adult males who have or had lung cancer, and see if there are any common factors. This involves selecting a historical sample. There are obvious difficulties with these retrospective studies. The researcher might omit or not recognize some common factors. It is difficult to detect if such a sample is biased or unbiased.

DEFINITION 2.5.10
A **HISTORICAL SAMPLE** is a sample selected of members of a population which are studied retrospectively to determine common features.

Another type of convenience sampling is called **JUDGMENT SAMPLING**. In this type of sampling, the researcher uses his or her own knowledge of the population or the knowledge of "experts" to select specific members of the population "to survey." Reports often will survey three or four "key" Senators or Representatives to find out how the Congress will vote on a particular issue. This would be an example of a judgment sample. The definition can be written as follows:

> **DEFINITION 2.5.11**
> **JUDGMENT SAMPLING** is a sampling technique in which a person knowledgeable on the subject and population selects the members of the population which he or she believes are most representative of the population.

There are two observations that we need to make concerning samples and the sampling process. The first is that when we sample, there will be discrepancies between the sample results and results that would be obtained from the whole population. There are actually two possible sources of these errors. The first type of error is due to factors inherent in the sampling process, such as using too small a sample. This type of error is called a **SAMPLING ERROR**. Sampling errors can usually be identified and described in a clear and understandable manner. The second type of error is due to factors outside the sampling process, such as human error or the use of a faulty instrument. These **NONSAMPLING ERRORS** are sometimes harder to detect, and harder to describe.

Our final note on sampling involves sample size. The authors have previously alluded to the fact that usually the larger the sample size, the better it represents the population. How large a sample is good enough? There are ways of determining, given the situation, how a large a sample is needed.

SECTION 6 -- SUMMARY

This chapter presented some key terms and concepts regarding statistical design and data collection. The intent of statistics is to make an inference or judgment about a population. Since studying the entire population can be costly, if not impossible, you may have to resort to sampling procedures.

KEY TERMS TO KNOW

Nominal
Cardinal
Ordinal
Discrete
Continuous
Research Design
Sampling

Simple Random Sample
Systematic Random Sample
Stratified Sample
Cluster Sample
Convenience Sample
Historical Sample
Judgment Sample

CHAPTER 3
DATA PRESENTATION

SECTION 1 -- INTRODUCTION

Most of the time data does not present itself in a nice neat package. It is not enough to process data; an individual must know what it means and be able to convey that meaning to others. How can you do that? It is the task of the researcher to take the raw data obtained from the sampling process of research design and put it into a presentable and understandable format. The old adage "A picture is worth a thousand words [or numbers]" is very much applicable in statistics. Many people find pictorial representations of data more readable and understandable than charts and tables. In this chapter, we will introduce a number of techniques of presenting data. Enough conceptual information will be given so that the data could be organized, summarized, or presented manually. However, more often than not, a computer is available to do the "busy work" desired. Then the major tasks are to decide on what procedures are desirable and to interpret the computer printouts. Just as there are different sampling techniques applicable for different problems, and sometimes more than one technique applicable to a given problem, there are different forms of data presentations which should be used on different types of data. The techniques presented in this chapter include organization of data through sorting, frequency distributions, stem and leaf diagrams, and graphical methods.

SECTION 2 -- FREQUENCY DISTRIBUTIONS

Suppose there is a couple looking to buy a house near West Point, New York. They are attracted to the town of Cornwall-on-the-Hudson, New York. They look at the classified ads in the local newspaper, **The Cornwall LOCAL**. They find ads for houses with prices given in the following example.

EXAMPLE 3.2.1:

$132,000	$785,000	$435,000	$395,000	139,900	125,000
$300,000	$229,000	$225,000	$199,000	179,000	375,000
$195,000	$175,000	$160,000	$135,000	375,000	215,750
$375,000	$375,000	$135,000	$197,500	199,900	189,900
$115,000	$147,000	$290,000	$159,900	186,000	169,900
$134,500	$205,000	$116,000	$139,900	169,000	168,500
$155,900	$159,000	$159,000	$160,000		

It is difficult to get a perspective on housing costs with the data in this format, so listing the costs in order would give a better perspective on housing costs.

\$115,000	\$139,900	\$160,000	\$186,000	\$215,750	\$375,000
\$116,000	\$139,900	\$160,000	\$189,900	\$225,000	\$375,000
\$125,000	\$147,000	\$168,500	\$195,000	\$229,000	\$395,000
\$132,000	\$155,900	\$169,000	\$197,500	\$290,000	\$435,000
\$134,500	\$159,000	\$169,900	\$199,900	\$300,000	\$785,000
\$135,000	\$159,000	\$175,000	\$199,000	\$375,000	
\$135,000	\$159,900	\$179,000	\$205,000	\$375,000	

However, in looking at the list above, it is still long; there are some prices that occur more than once and at the same time a number of gaps in the prices. One way to illustrate the same information would be to list the prices from lowest to highest with the number of houses offered at each price. Although this represents some improvement, the list would still be cumbersome. The key question is "What can be done to condense the list but still convey the same information?" The most important item is the information, idea or concept, not the actual data itself. In this particular example, the price of a house is the interest of the couple. In real estate transactions, the advertised price many times is meant only to give a "ballpark" estimate. It is not a fixed price, but is subject to negotiation. Also, in dealing with hundreds of thousands of dollars, how significant is a difference of \$500? Therefore, in this example, the couple can compress the list, without sacrificing information, by grouping prices together that are "close." But, how close is "close?" It will depend upon each problem and the data under consideration. In looking at the list in Example 3.2.1, the couple decides between themselves that houses over \$250,000 are too expensive, and thus can be grouped together. The remaining houses will be grouped together in \$25,000 intervals. The couple comes up with the following table:

Price Interval	**No. of Houses**
\$100,000 up to \$125,000	2
\$125,000 up to \$150,000	8
\$150,000 up to \$175,000	9
\$175,000 up to \$200,000	8
\$200,000 up to \$225,000	3
\$225,000 up to \$250,000	2
\$250,000 and over	8

How should groups be formed? Intuitively, the groups must satisfy several conditions. The first condition is that the grouping must make sense within the problem. Items that are grouped together must somehow fit together. In this example, the couple decides that houses over \$250,000 are too expensive, so they could be grouped together. The second intuitive condition is that the grouping makes sense with the data. For these housing prices, \$25,000 intervals make sense. The interval lengths are neither too large nor too small. The length is a good range within which the couple thinks they can find the house of their dreams. The third condition is that the number of intervals should be appropriate to the problem. Again, the number of intervals should not be too many or too few. [One rule of thumb that some texts use is that the number of intervals should be no fewer than 5 or more than 20.] These conditions can be summarized in the following convention:

CONVENTION 3.2.1

In grouping data, the following conditions should apply:

1. Grouped items go together naturally.
2. The lengths of intervals should make sense.
3. The number of intervals should be appropriate, usually between 5 and 20.

In most cases involving a sizable set of data, however, a computer is used; it usually has a default setting to organize the data into intervals, but these settings can usually be changed to determine both the number of intervals and the width of each.

For instance, when one statistical software package analyzes the housing price data above, the following default frequency distribution is produced:

Class Interval	Frequency	Percent Frequency
0 - 77,000	0	0
$78,000 - 155,000	10	25
$156,000 - 233,000	21	52.5
$234,000 - 311,000	2	5
$312,000 - 389,000	4	10
$390,000 - 467,000	2	5
$468,000 - 545,000	0	0
$546,000 - 623,000	0	0
$624,000 - 701,000	0	0
$702,000 - 770,000	0	0
$780,000 - 857,000	1	2.5
> $858,000	0	0

These intervals, however, do not seem appropriate for housing prices, so eight intervals of width 100 are selected with the following results:

Class Interval	Frequency	Percent Frequency
0 - $99,999	0	0.0
$100,000 - 199,999	26	65.0
$200,000 - 299,999	6	15.0
$300,000 - 399,999	6	15.0
$400,000 - 499,999	1	2.5
$500,000 - 599,999	0	0.0
$600,000 - 699,999	0	0.0
$700,000 - 799,999	1	2.5
> $800,000	0	0.0

You should be asking several questions at this point. What makes sense? What is appropriate? The answers to these questions really do depend upon the problem and the data under consideration. Most of the time some combination of number of intervals and length of intervals will suggest itself. Consider the following example:

EXAMPLE 3.2.2: The following are 33 scores from a final exam in a statistics class:

88	85	68	63	67	83
75	54	87	83	65	70
68	66	67	92	51	76
78	86	68	62	82	70
75	44	55	42	96	77
77	73	81			

Since these are test scores, they are percentages and could potentially range from 0 to 100. Thus 5 intervals of width 20 or 10 intervals of width 10 could be set up. However, the scores actually run from a low of 42 to a high of 96, so one could divide the data into 5 intervals of length 11, or 11 intervals of length 5. One can "justify" either approach if the context did not suggest another approach. Many times instructors will set up a grading scale prior to the test. This scale may be based on years of experience with similar classes. Suppose this particular instructor announced prior to the exam that the scale would be the following:

Grade	Interval
A	90 to 100
B	80 to 89
C	70 to 79
D	60 to 69
F	Below 60

In this particular case, grouping the grades according to this scale makes sense. Thus, the grouping must be appropriate to the problem and the data, and many times there is more than one "correct" grouping.

Having considered a set of data and grouped the items into appropriate classes, what do you do next? Consider the data on final examination grades, EXAMPLE 3.2.2. If the instructor chose to group the grades according to the preannounced grading scale, the distribution of grades in the class for the final exam would be the following:

EXAM GRADE	GRADE INTERVALS	FREQUENCY
A	90 to 100	2
B	80 to 89	8
C	70 to 79	9
D	60 to 69	9
F	Below 60	5

This distribution is known as a **FREQUENCY DISTRIBUTION**. Frequency distributions may be defined as follows:

> **DEFINITION 3.2.1**
> A **FREQUENCY DISTRIBUTION** of a variable is a description or listing of categories or classes of scores and the number of observations in each category or class.

There are certain conditions which are inherent in any frequency distribution. In EXAMPLE 3.2.2, you will note that each grade earned by some student in the class is included in the distribution. All grades are exhausted, and the distribution is said to be **EXHAUSTIVE**. Not only are all grades earned by students included, each grade is included in one and only one interval. The intervals are **MUTUALLY EXCLUSIVE**. The intervals also cover all grades that students could possibly earn. The intervals are said to be **INCLUSIVE**.

In some problems, the class intervals will be given or suggested by the problem itself. In other cases, the number and size of class intervals will need to be selected. If done on a computer, the data will automatically be put in classes, but most software allows for editing.

In working with distributions, sometimes it is more appropriate to consider proportions of data points than the actual number of points. The same information can be conveyed through a distribution known as the **RELATIVE FREQUENCY**, or sometimes also called the **PERCENT FREQUENCY**. To construct a relative frequency, one would take the total number of data points and calculate the proportion of grades in each interval.

DEFINITION 3.2.2
A **RELATIVE FREQUENCY DISTRIBUTION** of a variable is a description or listing of categories or classes of scores and the proportion of data points in each category or class.

For EXAMPLE 3.2.2, the relative frequency distribution would look like the following:

EXAM GRADE	GRADE INTERVAL	FREQUENCY	RELATIVE FREQUENCY
A	90 to 100	2	0.061 (2/33)
B	80 to below 90	8	0.242 (8/33)
C	70 to below 80	9	0.273 (9/33)
D	60 to below 70	9	0.273 (9/33)
E	Below 60	5	0.152 (5/33)
	TOTALS	33	1.001 (33/33)

What happened? The relative frequencies did not add up to 1.000, as expected. Since the total number of data points is 33, the individual decimal fractions are "not exact." They are "rounded off." For example, 2/33 = 0.060606.... Rounded off to three decimal places, 2/33 becomes .061. Therefore, the slight discrepancy is due to "round off error."

In working with distributions, there will be times when instead of considering just the frequencies of individual intervals we will need to look at the total number of data points up to and including a given interval. This distribution is known as the **CUMULATIVE FREQUENCY**. On occasions we will also find it useful to look at the cumulative percentages (also called the relative frequencies) and the **CUMULATIVE PERCENTAGE DISTRIBUTION**.

DEFINITION 3.2.3
A **CUMULATIVE FREQUENCY DISTRIBUTION** of a variable is a description or listing of categories or classes of scores and the total number of data points included in all categories up to and including a given category or class.

DEFINITION 3.2.4
A **CUMULATIVE RELATIVE OR CUMULATIVE PERCENTAGE DISTRIBUTION** of a variable is a description or listing of categories or classes of scores and the percentage of the total number of data points included in all categories up to and including a given category or class.

For EXAMPLE 3.2.2, the cumulative frequency and cumulative percentage distributions would look like the following:

EXAM GRADE	GRADE INTERVAL	FREQUENCY	CUMULATIVE FREQUENCY	CUMULATIVE PERCENTAGE
F	Below 60	5	5	15.15%
D	60 to below 70	9	14	42.42%
C	70 to below 80	9	23	69.70%
B	80 to below 90	8	31	93.94%
A	90 to 100	2	33	100.00%

Note that we "switched" the order from "A to F" to "F to A". We did this because intuitively when we speak of cumulative grades we go from the "bottom" up instead of the "top" down. In working with general distributions we can go either way, and in some case we will want to look at one direction for some applications and the other direction for other applications.

SECTION 3 -- GRAPHS

The use of a tabular format for distributions is not the only way to present data. Some people find it easier to understand data if there is a visual presentation of the information in a chart or graph. There are many different graphical or pictorial formats. Some are one dimensional; others are multidimensional. With the advent of computer graphics packages, it is possible to "paint a picture" to tell your story. One of the simplest forms is a two dimensional plot which takes its name from botany. Recall the test grades from EXAMPLE 3.2.2:

88	85	68	63	67	83
75	54	87	83	65	70
68	66	67	92	51	76
78	86	68	62	82	70
75	44	55	42	96	77
77	73	81			

Suppose we were to divide each of the grades into two parts: the tens digit, which we will call the **stem**, and the units digit, which we will call the **leaf**. The first score of 88 would be divided as follows:

STEM	LEAF
8	8

If we were to list the scores from the highest to the lowest, and combine all scores which had the same stem, the graphical representation would look like the following:

DIAGRAM 3.3.1:

STEM	LEAF
9	6 2
8	8 7 6 5 3 3 2 1
7	8 7 7 5 5 5 3 0 0
6	8 8 8 7 7 6 5 3 2
5	5 4 1
4	4 2

Stem and leaf graphs or displays are among the simplest of the graphical representations because each of the original data points or measurements is "visible." This format shows the data points in ascending or descending order, and for data sets which are not too large, it is fairly easy to construct. For larger data sets, most of the statistical software packages contain stem and leaf generators.

Another simple graphical representation of data is a two-dimensional graph, called a **BAR CHART**, in which vertical bars are placed along a horizontal axis. The vertical bars represent the various categorical groups in the data, while the height of the vertical bars represents the frequency of the given group. Recall the data from Example 3.2.2 in grouped form:

EXAM GRADE	GRADE INTERVAL	FREQUENCY
A	90 to 100	2
B	80 to below 90	8
C	70 to below 80	9
D	60 to below 70	9
F	Below 60	5

If we were to let each letter grade be represented by a vertical bar, we could plot this data as follows:

DIAGRAM 3.3.2:

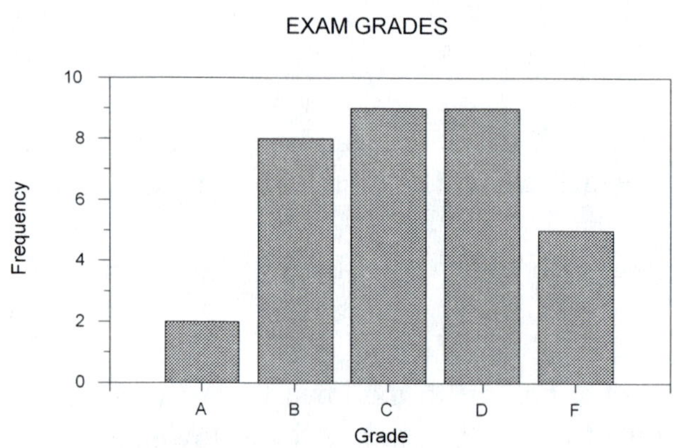

Or, the data could be represented numerically and on a 3-D graph as follows:

DIAGRAM 3.3.3:

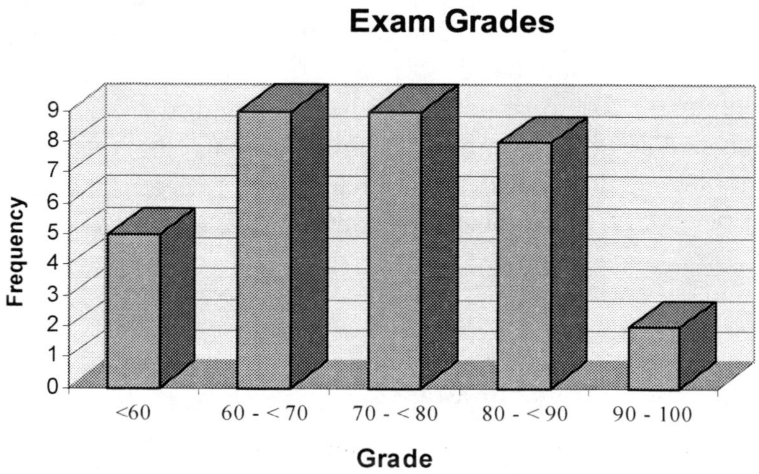

In both DIAGRAM 3.3.2 and DIAGRAM 3.3.3, the bars are disconnected from each other which is usually called a bar graph. Sometimes computer packages call this a histogram. Traditionally the difference between a bar graph and a histogram is that the bars in a bar graph do not touch (because the data is discrete/categorical) and in a histogram they do (because the data is continuous/interval or ratio). These distinctions are often blurred, however, in computer programs.

If we treat this data as continuous data, can we take advantage of some of the properties of continuous data. One way to do so is to transform the bar graph into a line graph. To do this, let us pick a representative from each of the groups. One can pick any numerical grade in the group to represent the group. However, one reasonable choice would be the center point of the interval. Computers usually have a default choice as illustrated in the following line graph for the grades:

DIAGRAM 3.3.4

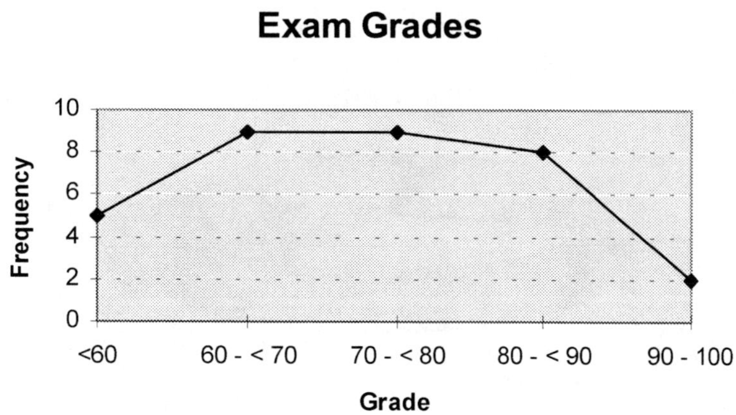

This graph is called a **FREQUENCY POLYGON**. In this format, height no longer represents frequency. Frequency is now represented by area under the curve, and thus we can use continuous methods, such as the calculus, to work with these graphs. We will not introduce these methods into this course, but you should be aware of their existence and their potential.

There are many other graphical formats. One particularly useful format is known as a **PIE CHART**. It is called this because the basic format is a circle with wedge-shaped pieces representing the various categories. To construct a pie chart by hand, we must first determine the center angle of each pie-shaped sector. We can do this by using the relative frequency and multiplying by 360, the number of degrees in a complete circle. If you have a good software package this is usually not necessary, since most will contain a pie chart generator, as illustrated next for the exam grades:

DIAGRAM 3.3.5

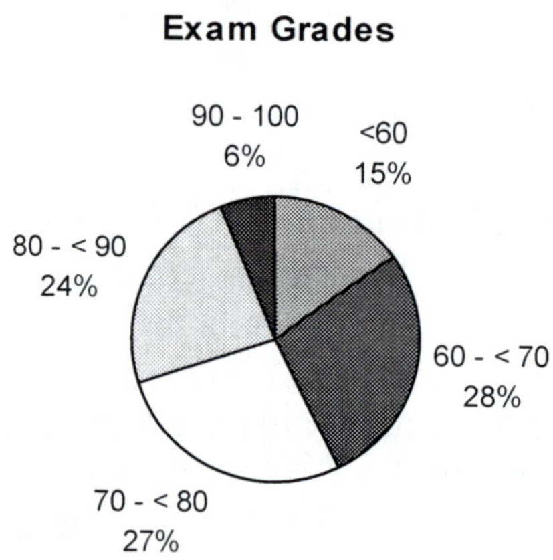

Or, the pie chart can be represented three-dimensionally:

DIAGRAM 3.3.6

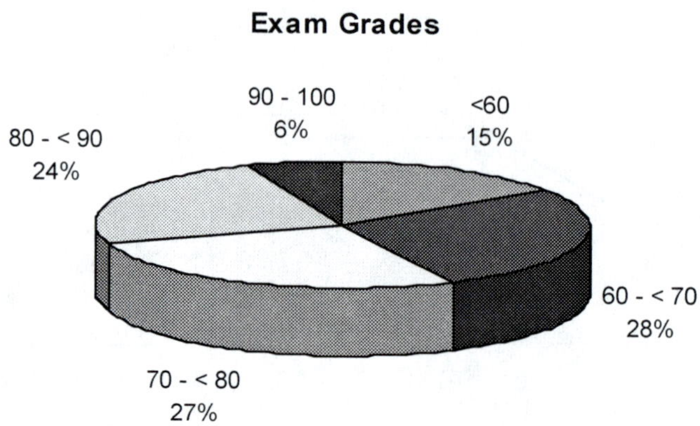

Suppose we were to consider graphing the cumulative frequency distribution for EXAMPLE 3.2.2 that we constructed earlier:

EXAM GRADE	GRADE INTERVAL	FREQUENCY	CUMULATIVE FREQUENCY	CUMULATIVE PERCENTAGE
F	Below 60	5	5	15.15%
D	60 to below 70	9	14	42.42%
C	70 to below 80	9	23	69.70%
B	80 to below 90	8	31	93.94%
A	90 to 100	2	33	100.00%

The line graph of a cumulative frequency distribution is called an **OGIVE.** DIAGRAM 3.3.7 illustrates the ogive for the data in our example.

DIAGRAM 3.3.7:

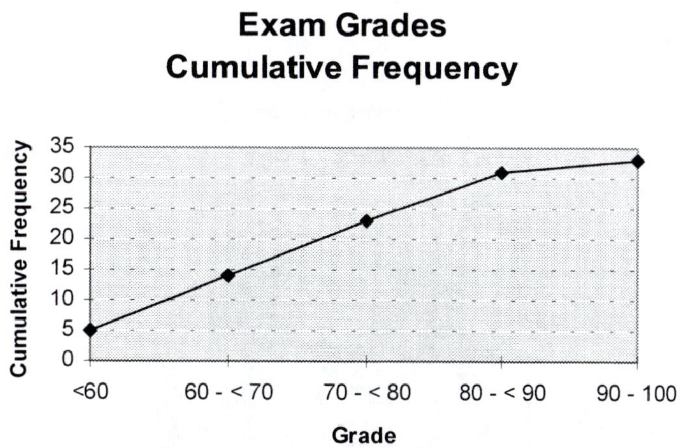

If the exam grades from Example 3.2.2 were from the first of two tests and the teacher wanted to compare the grades from Test 1 and Test 2, the two sets of grades could be contrasted both in tabular and graphic form:

EXAM GRADE	GRADE INTERVAL	TEST 1 FREQUENCY	TEST 2 FREQUENCY
F	Below 60	5	2
D	60 to below 70	9	5
C	70 to below 80	9	12
B	80 to below 90	8	10
A	90 to 100	2	4

DIAGRAM 3.3.8

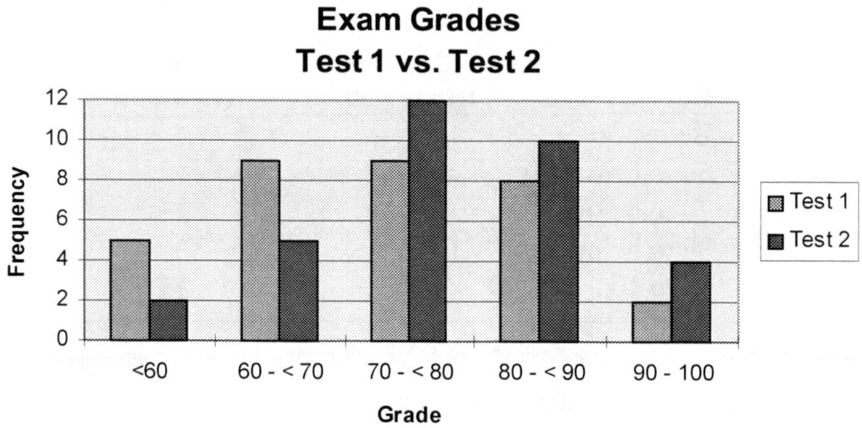

DIAGRAM 3.3.9

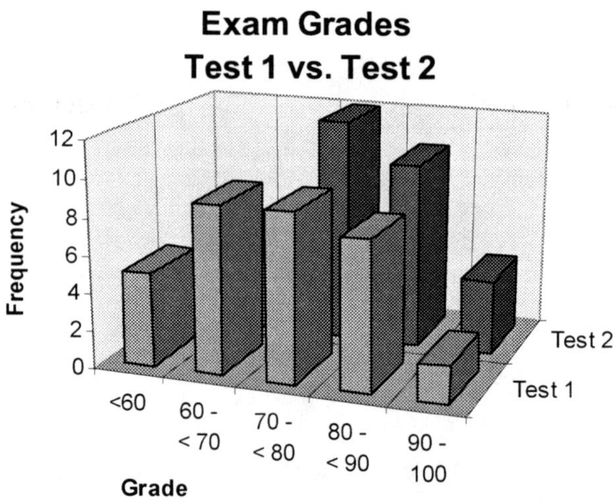

SECTION 4 -- ETHICS OF GRAPHS

Although a "picture [or graph] may be worth a thousand words," we must be certain that those words or ideas are really what we are trying to convey. In many ways it is easier to convey, purposely or inadvertently, the wrong information in graphs. Consider the following two graphs in DIAGRAM 3.4.1 and DIAGRAM 3.4.2:

DIAGRAM 3.4.1

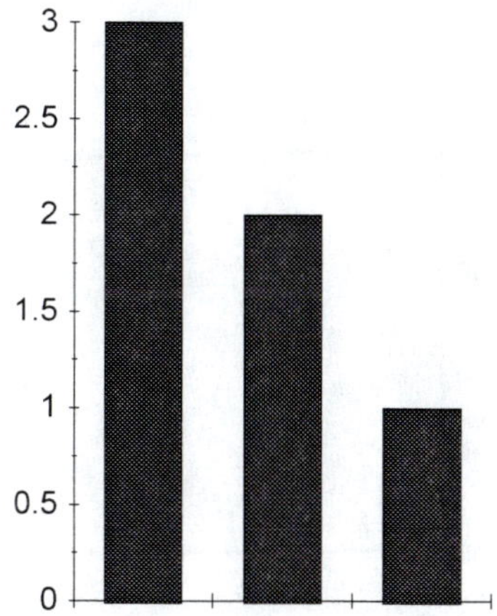

DIAGRAM 3.4.2

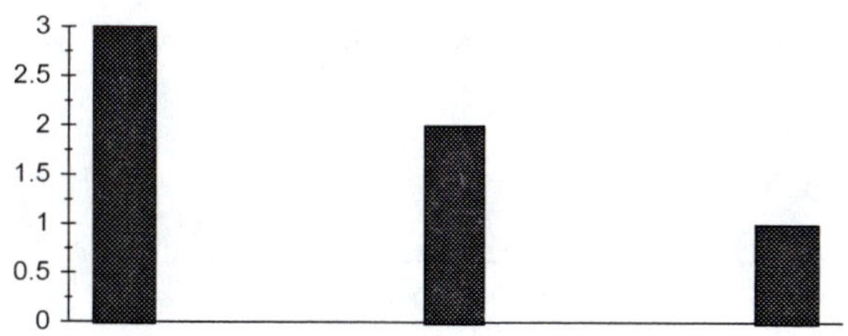

What impressions do you have of the data represented by the two graphs? Does it appear that there are large differences in the data represented by the graph in DIAGRAM 3.4.1, while there are much smaller differences in the data represented by the graph in DIAGRAM 3.4.2? However, if we look closely we see that the graphs represent the same data. How is that possible? The two graphs are using different vertical and horizontal scales. The three bars have heights of 3, 2 and 1, respectively. By shrinking or expanding the scale, we can produce significantly different impressions.

Different people looking at the same graph may have different impressions of what the graph conveys. We will never be able to eliminate that completely. Consider graphing the exam scores data from Example 3.2.2. The axes of the following two graphs have basically the same height and width, yet with different scales they give very different impressions:

DIAGRAM 3.4.3

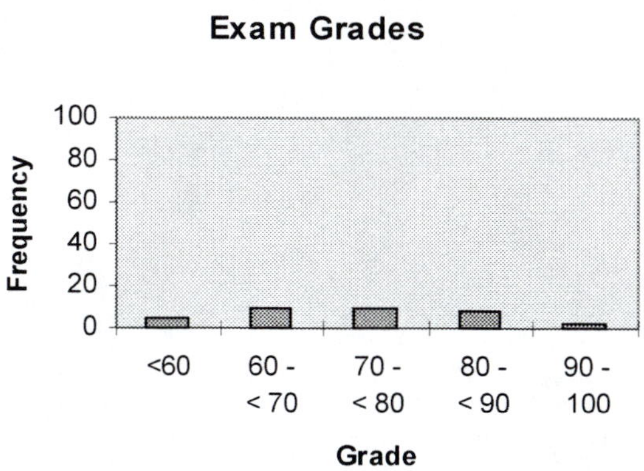

DIAGRAM 3.4.4

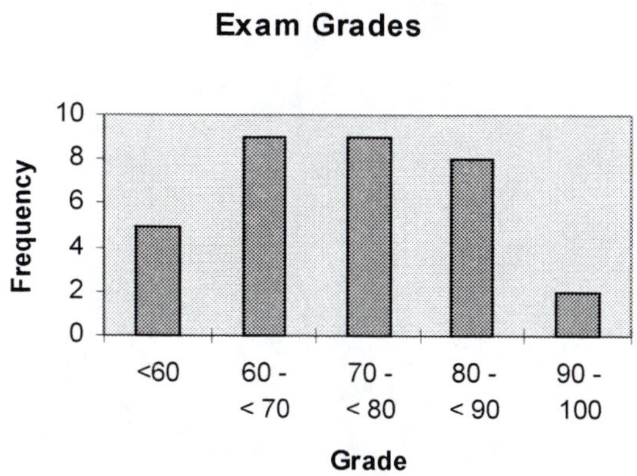

A head line in the newspaper for DIAGRAM 3.4.5 might read **Dow Jones plummets!**, while the head line for DIAGRAM 3.4.6 might be **Moderate Decline in Dow Jones**.

DIAGRAM 3.4.5

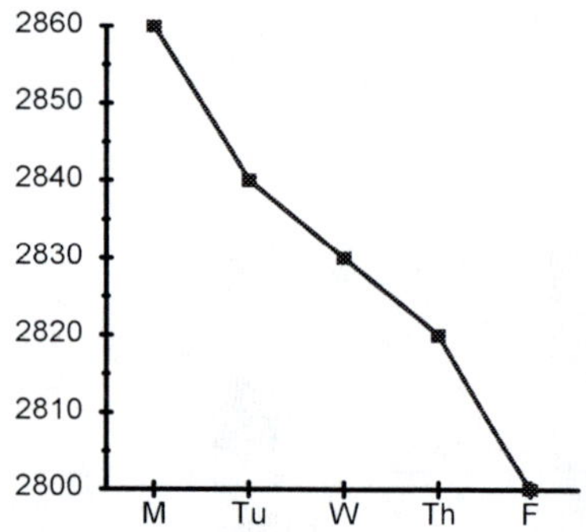

DIAGRAM 3.4.6

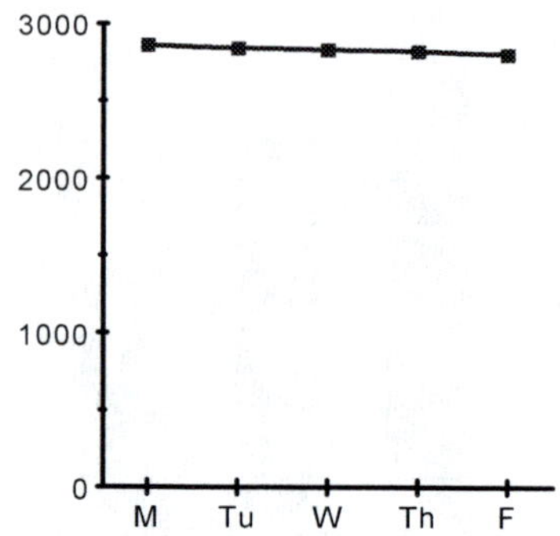

But if we look carefully at the two graphs, we see that they represent the same data. How can we avoid this problem? One way to help avoid the problem is to observe the convention that the initial entry on the vertical axis should be a zero frequency. If we were to follow this convention, we would not use DIAGRAM 3.4.5, sometimes called an **OH BOY CHART OR GRAPH** (for obvious reasons). This convention is not without its problems.

One of the main difficulties can be seen in DIAGRAM 3.4.6. If you were a newspaper publisher, you would not like to use DIAGRAM 3.4.6 in your paper. Why? There is too much "white space." There are no data points between 0 and 2800. Thus, all that space is opened. That takes up space in your newspaper which could be used for more of the article (or for advertisements, which would bring in revenue). How can we satisfy both concerns? The following example suggests a possibility.

DIAGRAM 3.4.7

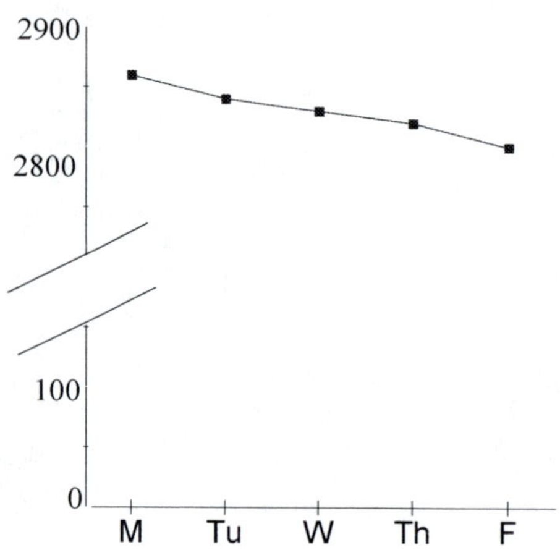

You can avoid many problems by always (or almost always) using these conventions. Unfortunately, they will not cover every situation. As in many areas of statistics, the key to drawing good graphs and charts is **common sense**. If you can honestly say that the graph or chart accurately conveys the data without bias, then most likely the people who will read the graph or chart will accurately interpret them.

SECTION 5 -- SUMMARY

In this chapter some key terms and concepts regarding data presentation are introduced. The tally method was introduced to condense the data but not lose any of the important information. Frequency distributions were introduced next. From frequency distributions, graphs were introduced. The simplest form of a graph was the stem and leaf display. Bar charts, histograms, frequency polygons, ogives, pie charts and high/low graphs were also discussed. Ethical considerations in using graphs were also addressed.

KEY TERMS TO KNOW

Tally
Class interval
Frequency distribution
Relative frequency distribution
Stem and leaf display
Bar chart
Histogram
Frequency polygon
Ogive
Pie chart
High/Low graph
Three-Quarter High Rule
Oh Boy Chart

CHAPTER 4
DESCRIPTIVE STATISTICS

SECTION 1 - INTRODUCTION

Did you hear the joke about the statistician who had his head in a heated oven and his feet in a tub of ice? On the average, he felt fine. What does the word "average" mean? Students talk about having a B+ average in statistics. Advertisers speak about the average American family having 1.5 children. Teachers refer to some students as "average students." In how many different ways are we using the word "average?"

Even from the examples cited above, we can see that the term "average" is used in at least three different ways. In order to avoid confusion, we will use a different term to refer to each of the different types of averages. But these averages will all have one thing in common. They are a single number or statement used to describe a group or class of objects.

When students refer to their average in a given class, they use one number or letter grade to describe all of their grades in that class. When advertisers speak about the "average" family, they try to describe those characteristics which are common to many, if not most, families. When teachers refer to "average" students, they speak about students whose demonstrated abilities would be "close to the middle" of the group.

The key here is the phrase "close to the middle." What do we mean by "center?" When we speak of an average, we speak about a description of the center of a group or class. In the next section of this chapter, we will discuss five different centers. But the center of a distribution does not completely describe the distribution. Intuitively, we know that it will be very difficult to describe a distribution with a single number or value. Consider the following example of three groups of student test results:

EXAMPLE 4.1.1:	**Group 1**	**Group 2**	**Group 3**
	60	69	80
	70	74	80
	70	76	80
	80	79	80
	80	80	80
	80	80	80
	80	80	80
	80	82	80
	90	84	80
	90	87	80
	100	89	80
Total	880	880	880
"Average"	80	80	80

If we were to look at just the "average" of the three distributions, there would be no difference. But looking at the distributions themselves, we can see that there are differences. All of the scores in Group 3 are identical. The scores in Group 2 are somewhat spread out, but the scores in Group 1 are even more widely scattered. This attribute is known as the variability or dispersion of the distribution. In the third section of this chapter, we will look at three measures of dispersion -- the range, variance and standard deviation.

In Section 4, instead of attempting to describe the whole distribution, we will focus on individual points. How can data points be described in relation to each other? How can data points in two different distributions be compared? Percentiles and the standard score are two ways of describing measures of location within a distribution. In Section 5, we look at several attempts to quantitatively or pictorially describe distributions concisely, but still in a fairly complete manner.

We will explain and illustrate these descriptive measures and show the simple calculations to obtain them, since we believe this helps to understand the concepts. However, in most instances the computer would be used to obtain these descriptive measures.

SECTION 2 - MEASURES OF CENTRAL TENDENCY

In the Introduction, we asked what was meant by "center" of a distribution. Although we do not have a precise definition of center, we do have an intuitive idea of what we mean. With that intuitive idea in mind, we can make the following definition:

DEFINITION 4.2.1
A **MEASURE OF CENTRAL TENDENCY** is a single number or phrase which describes the center of a distribution, class or group of objects.

Suppose a student earned grades of 70, 73, 65, 82 and 90 on five quizzes in a statistics class. The student could say that her average in the class was 76. What would she mean? How did she arrive at that result? We can answer the second question first. She obtained the result of 76 by adding the quiz scores together [70 + 73 + 65 + 82 + 90 = 380] and dividing by the number of quizzes [380/5 = 76]. This process is often referred to as averaging or finding the arithmetic mean of a set of numbers. We will simply refer to this as the **MEAN** of the set of numbers, and it gives us the following definition:

DEFINITION 4.2.2
The **MEAN** of a distribution or set of numbers is the arithmetic average of the data points, i.e., the sum of the numbers divided by the total number of data points.

The following example gives another illustration of finding the mean of a group of numbers:

EXAMPLE 4.2.1: Find the **average or mean weight** of the players of a high school basketball team, if the players' weights are 123, 137, 105, 175, 159, 180, 188, 195, 230, 208.

SOLUTION:

First add the weights: 123+137+105+175+159+180+188+195+230+208 = 1700.

Then divide the total by the number of players: 1700/10 = 170.

The mean weight is 170.

There is a shorthand notation which we can use to "simplify" the definition of mean. The Greek letter Σ (capital sigma) denotes the summation of a set of values. Thus the symbolic expression Σx means "sum all the values that the variable x can assume." Therefore, the formula of the mean of a distribution can be written as

$$mean = \frac{\sum x}{n} = \frac{x_1 + x_2 + \ldots x_n}{n}$$

where n is the number of data points.

While we are discussing notation, this would be a good place to discuss a convention which we will use throughout the remainder of this text. At times we will work with both samples and populations. If we are working with a sample taken from a larger population, we must be able to distinguish between the mean of the sample from the mean of the population. We will use the Greek letter μ (lower case mu) to denote the mean of a population and the symbol $\bar{x}$ (read x-bar) to denote the mean of a sample. In general, we will use Greek letters to denote characteristics of a population and English letters to denote characteristics of a sample.

MEAN: What does it mean?

We have answered the question, "How do you find a mean?" We still need to answer the question concerning the meaning of the term. One way to approach the meaning of the term "mean" is to consider the following: If we were to take the quiz score exam above and subtract the mean score (76) from each of the individual scores we would obtain the following set of scores: -6, 3, -11,6, 14. What happens when we add these scores: (-6) + (-3) + (-11) + 6 + 14 = 0. The sum of the differences from each data point and the mean is zero. We can think of this physically as the "balancing point" of the distribution. If we were to think of the picture of a seesaw with the scores at various points on the board, the mean is the place where we would have to place the fulcrum to balance the board. Note that although the board balances, it does not imply that there are an equal number of points on each side of the balance points. In fact, our example has three on one side and two on the other. What do we mean when we say that the "weights" are equally distributed from the "center."

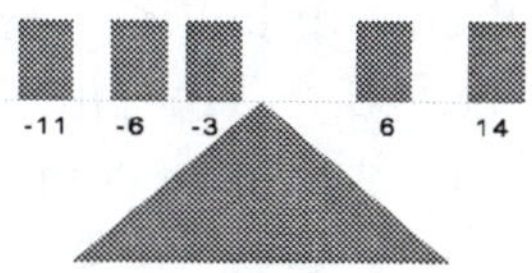

Suppose you were interested in the "average" family income in Westchester County, New York. Could you realistically hope to find the mean of the population? Since such a task would be impossible, we would need to "estimate" the mean, using a sample. The difficulty in this will be finding a representative sample. Westchester County borders on New York City. It is almost three different counties in one. It has several urban centers with the associated pockets of poverty. It has a number of "bedroom communities" from which the upper and upper-middle class residents commute to their jobs in New York City. It also has a number of middle class suburban communities with associated light and service industries.

EXAMPLE 4.2.2: Suppose we obtained the following sample of family incomes:

$2,600,000	$1,250,000	$750,000	$485,000
$ 375,000	$ 115,000	$ 85,000	$ 78,000
$ 40,000	$ 30,000	$ 29,000	$ 26,000
$ 19,500	$ 14,000	$ 12,000	$ 12,000
$ 12,000	$ 11,800	$ 10,900	$ 6,700

The number of data points is n = 20 and the sum of the data points is $ 5,961,900. Thus, the mean of this sample would be:

$$\bar{x} = \frac{\sum x}{n} = \frac{\$5,961,900}{20} = \$298,095$$

Thus, the "average" family income is almost $300,000. Does that one number give an "accurate" picture of the family income situation in the county? It does indicate that there is great wealth in the county. It does not indicate that there is also poverty.

If we look at the mean of this example, we see that there are five data points above the mean and fifteen below. This would say that the mean is not completely "representative" of the family income situation. What would be more representative. One answer would be the "half-way" mark, the score (or point) at which half of the scores are above and half are below.

In this example, there are 20 scores. We would be looking for the point at which 10 scores are above and 10 below. Since the scores are arranged in decreasing order, we can count down 10 scores to get to the income of $30,000. If we were to count up 10 scores from the bottom, we would get to the income of $29,000. Thus, the halfway mark will be somewhere between $29,000 and $30,000. We will by convention take the point halfway between those two scores. Therefore, in this example the halfway mark is $29,500. This income figure intuitively seems more representative of the family income situation within the county. It is not skewed by the extreme values by which the mean was. This type of "average" will be called the median.

DEFINITION 4.2.3
The **MEDIAN** of a distribution or set of numbers is the point at which half of the scores are above and half below.

Consider the following example:

EXAMPLE 4.2.3: Find the median score of the following quiz results: 70, 73, 65, 82 and 90. We first need to arrange these scores in ascending (or descending) order: 65, 70, 73, 82, 90. Since there are five scores, we can find the middle score. In this case it is 73. Therefore, the median quiz score is 73.

Note the slight difference between the two examples. In the first example of the family incomes, there are an even number of data points. In the second example of the quiz scores, there are an odd number of data points. For a problem with an odd number of data points, we are able to select a score within the distribution for which there are the same number of scores above as there are below. For a problem with an even number of scores, we choose a point halfway between two scores, so that there are the same number of scores above the median as below.

There are occasions when the mean and median are not the most appropriate measures of central tendency. Suppose you are the manager of a shoe store. You are responsible for ordering a sufficient supply of shoes in the appropriate sizes to satisfy your customers. Can you use the "average" shoe size of people to help you? Does it make any sense to talk about the mean shoe size of a group of people? Not really. What about the median shoe size? It does make more sense than mean shoe size, but it still does not tell you much about what sizes to stock.

Is there any type of "average" that does? Can we talk about the average shoe size for males being 10D? There is a sense in which we can. What we are saying is that the most used or most popular shoe size is 10D. We will call this type of average the mode.

DEFINITION 4.2.4
The **MODE** of a distribution or set of numbers is the score or data point which occurs most frequently.

The following examples illustrate various aspects of modes, as well as how to determine modes of different types of distributions:

EXAMPLE 4.2.4: Given the sample of family incomes from Westchester County which we used before, find the mode. The distribution is as follows:

$2,600,000	$1,250,000	$750,000	$485,000
$ 375,000	$ 115,000	$ 85,000	$ 78,000
$ 40,000	$ 30,000	$ 29,000	$ 26,000
$ 19,500	$ 14,000	$ 12,000	$ 12,000
$ 12,000	$ 11,800	$ 10,900	$ 6,700

Since the value $12,000 occurs 3 times, and no other value occurs more than once, the mode of this sample is $12,000.

EXAMPLE 4.2.5: Given the following set of quiz scores (graded on a scale of 0 to 10) for a statistics class, determine the mode for the class.

4	5	7	9	10	4	9	10	8	3
10	6	7	10	8	6	3	7	7	

In this class of 19 students, we find the following distribution of scores:

Score	Number of Students
3	2
4	2
5	1
6	2
7	4
8	2
9	2
10	4
TOTAL	19

In this distribution, we see that there are two scores, 7 and 10, which occur most frequently; each of these scores occurs 4 times. No other score occurs more than twice. In this situation where there are two scores which occur most frequently, we say that the distribution is bimodal. In cases where more than two values are the "most frequent," we say that the distribution is multimodal. In cases where no score is repeated, we say that the distribution has no mode.

WHICH IS THE BEST AVERAGE?

In our example of the family incomes from Westchester County, we calculated the mean ($298,095), the median (29,500) and the mode ($12,000). Which is the best average? ***IT DEPENDS!***

The measure of central tendency that you use should be appropriate to the problem and to the desired result. The mean in this example gives an accurate estimate of the very large gross income of the county. The median in this example gives a fair picture of the economic class structure of the county. The mode in this example tells us that there is a large pocket of relative poverty within a wealthy area.

In working with examples like this, one measure may not tell us enough of the information that we want or need to know. However, in combination with others measures, we can get a reasonable picture of the situation.

We noted in our discussion of the median the fact that the mean was highly susceptible to "outliers" or extreme values, those scores far from the mean and far from the remainder of the scores. The median was one way to reduce the influence of the outliers. Another way is known as the trim mean. As the name implies, it is a mean from which something has been trimmed or eliminated. If we wish to reduce the influence of the extreme values, one way would be to eliminate some of the most extreme values. Thus, to find the trim mean of a distribution, we eliminate five percent of the extreme values on each side of the distribution, and calculate the mean of the remaining values.

EXAMPLE 4.2.6: The Jones Co. employs 13 people whose weekly salaries are:

$600	$790	$750	$720	$820
$820	$749	$900	$550	$1350
$820	$770	$920		

The mean of these 13 salaries is $812.23. To find the trim mean, we first need to determine how many values to eliminate. Five percent of 13 is 0.65. Since we do no want to eliminate a partial value, we would need to eliminate 1 value on each side. This means we would eliminate the values of $1350 and $550 and calculate the mean of the remaining scores. This value would be $787.18. The trim mean should be closer to the median than the mean is. In this example, the median is $790, which is very close to the value of the trim mean.

The preceding example leads us to the following definition:

> **DEFINITION 4.2.5**
> The **TRIM MEAN** is the mean value of the distribution obtained by eliminating the same percentage (usually 5% or 10% of extreme values on both sides of the distribution.

There is one additional measure of central tendency which we need to introduce. If we have defined a measure of central tendency to be a number which describes the "center" of a distribution, then a "center" of the distribution is the point halfway between the maximum value and the minimum value. This point is called the **MIDRANGE**.

> **DEFINITION 4.2.6**
> The **MIDRANGE** of a distribution or set of numbers is the arithmetic mean of the maximum and minimum values of the distribution, i.e., the point halfway between the maximum and minimum values.

The following example will help to illustrate the interrelationship among the five measures of central tendency.

EXAMPLE 4.2.7: If we return to the Jones Co. with its 13 employees whose weekly salaries are:

\$600	\$790	\$750	\$720	\$820
\$820	\$749	\$900	\$550	\$1350
\$820	\$770	\$920		

From EXAMPLE 4.2.6 we obtained:

mean	=	\$ 812.23
median	=	\$ 790.00
trim mean	=	\$ 787.18

To determine the midrange, we need to identify the maximum and minimum values of the distribution. The maximum value is \$ 1350 and the minimum value is \$ 550. The arithmetic mean of these two values is \$ 950.00. Therefore,

midrange = \$ 950.00

Since the only data point which occurs more than once is \$ 820, we have the following:

mode = \$ 820.00

Since the trim mean eliminates the outliers, it will be closer to the median than will the mean. Since the midrange is solely dependent upon the extreme values, it can be more "volatile" than the other measures. In this example, it is the "most extreme" of the measures of central tendency.

In a small distribution (few data points), the mode may not have the significance that it will have in a large distribution (many data points). In a large distribution the median will "fall between" the mean and the mode. This distribution is too small with too many extreme values for the median property to hold.

SECTION 3 - MEASURES OF DISPERSION

In the introductory section of this chapter, we introduced the attribute of variability or dispersion of a distribution or set of numbers. In this section we will be looking at several measures of dispersion.

The simplest measure of dispersion is how far the distribution extends. With this measure, we look just at the difference between the highest value and the lowest value. We will call this measure the range.

DEFINITION 4.3.1

The **RANGE** of a distribution is the numerical difference between the maximum (greatest) and the minimum (least) value in the distribution.

If we were to calculate the range for the three distributions in EXAMPLE 4.1.1, we would find that the range for Group A is 40 (100 - 60). For Group B it is 20 (89 - 69) and for C it is 0 (80 - 80). If we were just given these three values of the range and no other information about the distributions, we could conclude that there were indeed differences in the distributions. If the ranges are different, the distributions are different.

However, differences in the value of the range still do not tell us everything about the distribution. Consider the following example:

EXAMPLE 4.3.1

Group B	Group D
69	71
74	71
76	72
78	72
80	80
80	80
80	80
82	88
84	88
87	89
89	89

For Groups B and D, the mean, median and mode are all 80. The range of Group B is 20 (89 - 69), while the range of D is 18 (89 - 71). For these two groups, the range of D is less than the range of B. Intuitively, however, Group B is not as spread out as Group D.

The range is very simple to calculate. It only uses two values from the distribution--the maximum and the minimum. But its simplicity can be a problem. It does not really tell us anything about how the data points are distributed within the distribution. In order to learn anything about how the data points are distributed, we must go to something more complicated--the variance.

One way to include all values within the distribution in our measure of dispersion is to somehow measure the "average" distance of all data points from the mean. What happens if we try this? Consider the following example:

> **EXAMPLE 4.3.2:** Consider the following set of data points:
>
> 2, 3, 4, 5, 5, 5, 6, 7, 8.
>
> The mean is 5. [Why?] If we were to measure the distance of each point from the mean we would have the following distribution:
>
> -3, -2, -1, 0, 0, 0, 1, 2, 3
>
> [Why?] If we were to calculate the mean of this distribution, it would turn out to be 0. In fact, this is a property of the mean. We discussed this earlier in this chapter. The mean is the "balancing point" of a distribution.

If this average is always 0, how can we talk about using this concept to measure the dispersion of a distribution? Note that in the distribution of distances from the mean, some of the distances are negative (i.e., some points are below the mean) and some of the distances are positive (i.e., some points are above the mean). However, when we talk about distance do we usually talk about negative distance? Do we say that the distance from Buffalo (NY) to Albany (NY) is 480 miles, while the distance from Albany to Buffalo is -480 miles? We usually say that the distance is 480 miles, no matter which direction we are going.

Thus we need to treat distances as positive numbers. The distribution of distances from the mean in this example becomes:

3, 2, 1, 0, 0, 0, 1, 2, 3.

The mean of this distribution is 1.33. [Why?] This value represents the "average" distance of each data point from the mean. This is a "good" measure in that it does involve every data point. Unfortunately, there are mathematical problems in working with absolute values.

How can we still use "positive" numbers, but eliminate the problems of using absolute value. There is one other way of guaranteeing that all numbers will be positive, or at least non-negative. We know that the square of any number will be greater than or equal to 0. So we could square each distance. The resultant distribution will be as follows:

9, 4, 1, 0, 0, 0, 1, 4, 9.

In addition to the fact that all the numbers are non-negative, there is one additional feature of this distribution. By squaring the numbers, the extreme values are accentuated. To calculate the mean of this distribution, we would add the numbers and divide by 9, the number of data points. In this example, it would give us 28/9 = 3.11. But since 3.11 is the "average" of the squared deviations from the mean, we need to take the square root to get back to the original units. The square root of 3.11 is 1.76. In other words, the data points are approximately an average of 1.76 units from the mean of 5. This measure of variability is called the **standard deviation**.

DEFINITION 4.3.2
The **STANDARD DEVIATION** of a distribution is the square root of the average deviation from the mean squared; intuitively, it is approximately the average distance of the data values from the mean of the distribution.

In practice, the standard deviation of a set of data is almost always determined on a computer. Therefore, we will not compute the standard deviation by "hand," but will give its formula at the end of the chapter for reference. The computer will also usually give the **variance** of the distribution, which is just the standard deviation squared, or the result of 3.11 from our example before we took the square root. It is useful for theoretical purposes in inferential statistics, but not as easily interpretable since it is not in the original units of the data.

Earlier in this chapter we noted the convention that we will use the Greek letter μ to represent a population mean and the English $\bar{x}$ to represent a sample mean. In general, we will use Greek letters to represent population parameters and English letters to represent sample statistics. In this case, we will use the Greek letter σ (lower case sigma) to denote the population standard deviation and the lower case English letter s to denote the sample standard deviation. There is a slight difference between the formulas (and thus answers) for the population and sample standard deviations due to theoretical issues. In most cases, we will be dealing with the sample standard deviation since usually we do not have the entire data set for the population. The symbolism is reviewed in the following convention:

CONVENTION 4.3.1 - LIST OF SYMBOLS

Σ	Denotes summation of a set of values
x	Is a variable used to represent the individual scores
n	Represents the total number of scores under consideration
$\bar{x}$	Denotes the mean of a sample
μ	Denotes the mean of a population
σ	Denotes the population standard deviation
s	Denotes the sample standard deviation

EXAMPLE 4.3.3: Find the standard deviation of the weights of the basketball players from EXAMPLE 4.2.1 (consider them as a population). The weights (in pounds) are:

123, 137, 105, 175, 159, 180, 188, 195, 230, 208

SOLUTION: By calculator or computer, the mean weight of the players is 170 and the population standard deviation is 37.10. In other words, the "center" of the weights is 170 pounds and the "spread" is 37.10 pounds.

EXAMPLE 4.3.4: Suppose we choose a sample of 16 students from an entering freshman class of 600. For these 16 students, their combined SAT scores are:

1100	1170	1230	1350	1090	1450	1250	1290
1180	1290	1380	1520	1080	1300	1270	1210

The sample mean is $\bar{x}$=1260 and the sample standard deviation is s=125.05 (done on the computer). In other words, the "center" SAT score is 1260 and the "average spread" from 1260 is approximately 125.05998 points.

When should we round and when should we not round? This is not an easy question to answer. The idea of **SIGNIFICANT FIGURES** is important and cannot be ignored. If we are working with measurements in **kilograms**, it does not make sense to report a statistic in **milligrams**. Generally speaking, standard deviations should be reported to two significant figures.

To complete this section on measures of dispersion, we ask the question, "Is there a relationship between the range and the standard deviation?" Mathematically, there is no formula relating the two measures. However, there are some guidelines which we can use to estimate the bounds on one measure given the other measure.

If we are working with a population, then it is possible to prove that the population standard deviation can never be more than one-half of the population range. By using the same arguments, it is also possible to show that for samples, the sample standard deviation can never be more than $1/\sqrt{2}$ (or approximately 71%) of the sample range. As the number of data points increases, whether we are dealing with a population or sample, the standard deviation tends to be a slightly smaller proportion of the range. The following convention puts bounds on the ratio of the standard deviation to the range.

CONVENTION 4.3.2
If $n \leq 100$, then the **STANDARD DEVIATION** is almost always between 1/6 and 1/2 of the **RANGE**.

Usually the range of our population or sample is easier to calculate or determine than the standard deviation. Knowing the range, the convention above gives us a "quick and dirty" estimate of the standard deviation of our population or sample. The following examples illustrate this:

EXAMPLE 4.3.5: In EXAMPLE 4.2.1, we introduced a basketball team whose members had the following weights: 123, 137, 105, 175, 159, 180, 188, 195, 230, 208. The range of weights for these individuals is

$$\text{RANGE} = 230 - 105 = 125$$

CONVENTION 4.3.2 says that the STANDARD DEVIATION should be between 1/6 and 1/2 of the range. Now 1/6 of 125 is 20.83 and 1/2 of 125 is 62.50. Thus our standard deviation σ should be between the following limits

$$20.83 \leq \sigma \leq 62.50$$

In EXAMPLE 4.3.3, we calculated the STANDARD DEVIATION of this population. We found $\sigma = 37.1$, which is within our limits.

EXAMPLE 4.3.6: In EXAMPLE 4.3.4 we introduced a sample of 16 students from an entering college freshmen class of 600 students. For these 16 students, their combined SAT scores were:

1100 1170 1230 1350 1090 1450 1250 1290
1180 1290 1380 1520 1080 1300 1270 1210

The range of this sample = 1450 - 1080 = 370

In EXAMPLE 4.3.4, we found the sample standard deviation to be $s = 125.05$

The ratio of the sample standard deviation to the range is given by

$$\frac{s}{range} = \frac{125.05}{370} = 0.34$$

which is within the limits of 1/6 to 1/2 of the range.

CONVENTION 4.3.2 is a convention. It is not an absolute rule. There might be situations when the standard deviation falls outside the limits of 1/6 to 1/2 of the range. However, these limits provide a good guideline against which to check the calculations of a standard deviation. If we have a standard deviation which falls outside these limits, we should recheck our calculations. There is a very good chance we have made an error in our calculations.

SECTION 4 - MEASURES OF LOCATION

Given a distribution or set of data, what have we done so far in this chapter? We have looked at **MEASURES OF CENTRAL TENDENCY**, i.e., "Where is the center of the distribution?" We have also looked at **MEASURES OF DISPERSION**, i.e., "How spread out is the distribution?" These are attempts at numerical descriptions of the entire distribution. Sometimes we are interested in the relative position of data points within the distribution, i.e., **MEASURES OF LOCATION**.

DEFINITION 4.4.1
A **MEASURE OF LOCATION** is a single value that describes the relative position of a data point within a distribution.

The most readily apparent measures of location of points within a set of numerical data are values which we have already used several times. We have used these measures of location in the calculation of the range and the midrange. We are talking about the extreme values, the highest and lowest values of the distribution. These extremes are usually referred to as the **MAXIMUM** and **MINIMUM**. If you will recall the range was the numerical difference between the maximum and minimum. The midrange was the arithmetic mean of the maximum and minimum. What are the highest and lowest values of a distribution? Intuitively, we know that all observations will be less than or equal to the maximum. Likewise, all observations will be greater than or equal to the minimum. We will use the suggestive notation $\mathbf{x_{max}}$ and $\mathbf{x_{min}}$, or **H**(high) and **L**(low), for the maximum and minimum respectively. We can formalize this discussion into the following definitions:

DEFINITION 4.4.2
The **MAXIMUM** of a set of observations is the largest observation within the set.

DEFINITION 4.4.3
The **MINIMUM** of a set of observations is the smallest observation within the set.

The following example illustrates the definitions above.

EXAMPLE 4.4.1: Given the weights of the basketball players in EXAMPLE 4.2.1, what are the maximum and minimum values of this set of data?

SOLUTION: The weights of the basketball players are:
123, 137, 105, 175, 159, 180, 188, 195, 230, 208.

The greatest weight is 230. The least weight is 105. Therefore, we have

$$H = x_{max} = 230 \qquad \text{and} \qquad L = x_{min} = 105$$

The maximum and minimum are easily explained and understood. However, they can be very misleading. If one observation (or just a few observations in a data set with many points) is very different from all the rest of the data points, then the max or min could give a false impression about the distribution. Recall our discussion about the family incomes in Westchester County, New York, in EXAMPLE 4.2.2. There was one annual income value of $2,600,000 in our sample. This value was more than twice any other value in the sample. This one value greatly influences our perception of family income in Westchester county.

Looking at the maximum and minimum values can be useful in another way. Errors can and do occur in the observation and recording of data. Instruments can be read incorrectly. Instruments can be wrong at times. Readings can be recorded incorrectly. It is a good idea to check the maximum and minimum values of any distribution to make certain that the data makes sense. Consider the following example:

EXAMPLE 4.4.2: Suppose you are a traffic engineer and are studying the traffic patterns at a city intersection. One of your researchers brings in the following set of data (cars per hour passing through intersection):

8 - 9 am	340
9 - 10 am	450
10 - 11 am	390
11 - 12 noon	430
12 - 1 pm	475
1 - 2 pm	297
2 - 3 pm	23
3 - 4 pm	310
4 - 5 pm	525

Now $x_{max} = 525$ and $x_{min} = 23$. What is wrong with this data? The minimum value is so far out of line with the rest of the data that the data should be called into question. More than likely in this case the value "23" is wrong. It might have been a mistake in the recording of data. The true value might have been "230" or "223". It might have been a surveyor error. Usually traffic counts are taken on a machine which has a reset button. The reset button might have been accidentally pushed near the end of the 2 to 3 pm hour. With a questionable minimum value such as this, the most prudent line of attack would be to send another researcher out to this corner for a recount.

The maximum and minimum located two specific points within a distribution. What happens if we are interested in comparing points within a distribution or comparing the relative position of points in two distributions. For example, how would you compare a score of 530 on the Verbal SAT with a score of 24 on the English ACT? The problem is that the two distributions have different scales or units of measure. Can we "fix" the distributions so that they both have the same scale? In fact, there are at least two methods which have been used extensively to "solve" this question.

The first solution makes use of the mean and standard deviation of the distribution. The following examples illustrates the "how" and suggests the "why" of this measure.

EXAMPLE 4.4.3: Suppose the mean height of American males is 70 inches with a standard deviation of 3.4 inches and the mean height of American females is 65 with a standard deviation of 2.8 inches. If we also assume that a brother and sister are 76 inches and 70 inches respectively, which is taller relative to others of their own gender?

SOLUTION: Obviously, the brother is 6 inches (76 minus 70) taller than his sister. However, that is not what was asked. We need to compare the brother against all males and the sister against all females. One possible way of looking at this problem would be to ask how much taller than the mean is each. The brother is 6 inches taller than the mean of males, while the sister is 5 inches taller than the mean of females. So the brother is further away from the mean of males than the sister is from the mean of females.

But that answer does not really tell us **WHERE** within the distribution of heights of males and females are the brother and sister. The standard deviation is a measure of how spread out the data points are within the distribution. We could **MEASURE** how far a given data point is from the mean in terms of the standard deviation, i.e., we divide the difference between the data point and the mean by the standard deviation. For the brother we have

$$\frac{76-70}{3.4}=\frac{6}{3.4}=1.76$$

For the sister, we have

$$\frac{70-65}{2.8}\frac{5}{2.8}=1.79$$

What do the 1.76 and 1.79 mean? First of all, what are the units of these numbers? These numbers are quotients. The units of the numerator are inches and the units of the denominator are also inches. The inches thus cancel out each other and the quotient is "unitless". Then, what are 1.76 and 1.79? How did we get these numbers? We took the differences of our data points from the respective means and divided by the respective standard deviations. It is in effect the number of standard deviations our data points are from the respective means. We could say that the units are **STANDARD DEVIATION UNITS**. In terms of standard deviation units, the sister is further away from the mean of females than her brother is from the mean of males.

In the example above, we introduced the concept of **STANDARD DEVIATION UNITS**. We can formalize this concept in the following definition:

DEFINITION 4.4.4
The **STANDARD SCORE** associated with a data point within a distribution is the number of standard deviations the data point is from the mean of the distribution.

The standard score for a given data point x from a population will be denoted by $\mathbf{z_x}$. From the definition we have the following computational formula for the standard scores from a population:

STANDARD SCORE (POPULATION):

$$z_x=\frac{x-\mu}{\sigma}$$

For standard scores from a sample, the formula would be:

STANDARD SCORE (SAMPLE; μ UNKNOWN):

$$z_x = \frac{x - \overline{x}}{s}$$

The subscript notation for standard scores is meant to help us identify at which data point we are looking. Most of the time it will not be necessary to use this notation. In fact, if there is no chance of confusion, the subscript x will be dropped.

The following example provides another example of how standard scores could be used to compare scores from two different distributions:

EXAMPLE 4.4.4: What is a "better" score: 630 on the Verbal SAT or 32 on the English ACT?

SOLUTION: Suppose the national mean for the Verbal portion of the SAT exam is 428 with a standard deviation of 108 and the national mean for the English portion of the ACT exam is 21.2 with a standard deviation of 5.3. What are the standard scores associated with these two test results?

$$z_{630} = \frac{630 - 428}{108} = \frac{202}{108} = 1.87$$

$$z_{32} = \frac{32 - 21.2}{5.3} = \frac{10.8}{5.3} = 2.03$$

Since $z_{32} > z_{630}$, the score of 32 on the English ACT is **BETTER** then the score of 630 on the Verbal SAT.

Most of you are familiar with the SAT and ACT exams. These two national standardized exams report scores two ways. They report what they call a **SCALED SCORE**. For each of the two parts of the SAT, the scaled scores range from 200 to 800. For each of the four parts of the ACT, the scaled scores range from 1 to 36. Each of the major testing firms also report their scores a second way. This second way is called a **PERCENTILE SCORE**. For the Verbal SAT, the scaled score of 630 would be reported as a percentile score of 97, or a 97th percentile. For the English ACT, the scaled score of 32 would be reported as a percentile score of 97 also.

The idea of a percentile score is generally stated as the percentage of the distribution which falls below the given score. This means that a Verbal SAT score of 630 is greater than the score for 97 percent of those individuals who took the exam. Does that mean that 3% of the test takers scored better than 630? Not necessarily! Consider the following example:

EXAMPLE 4.4.5: Reconsider the distribution of weights of the players on the high school basketball team introduced in EXAMPLE 4.2.1:

123, 137, 105, 175, 159, 180, 188, 195, 230, 208

What percentage of the weights are below 180? What percentage are above 180?

SOLUTION: To find the percentage of weights below 180, we could first rank the weights in ascending (or descending) order:

105, 123, 137, 159, 175, 180, 188, 195, 208, 230

How many weights are less than 180? There are 5 (105, 123, 137, 159 and 175). Since there are a total of 10 weights, the 5 weights below 180 represent 50% of the total number of weights.

How many weights are greater than 180? There are 4 (188, 195, 208, 230). These four scores represent 40% of the number of weights.

What is the "problem" in EXAMPLE 4.4.5? One possible problem is that there are only 10 data points. Thus each data point represents 10% of the scores. In the SAT example, each SAT score represents much less of the total distribution. In the case of SAT's, if 97% of the scores fall below a given SAT score, then there will be approximately 3% above that score. This will be the case for most large populations or samples. Because of this property, the use of percentiles may not be appropriate for small populations or samples. In any case we formalize our definition of percentile as follows:

DEFINITION 4.4.5
The **PERCENTILE RANK** of a score within a distribution is the percentage of scores within the distribution below the given score.

The following examples illustrate how we can calculate percentile ranks from raw scores and how we can determine raw scores given percentile ranks in various situations:

EXAMPLE 4.4.6: Recall the distribution of weights of high school soccer players from EXAMPLE 3.2.4:

160	155	147	200	200	148	150
155	158	135	145	135	130	145
140	155	131	155	140	120	140
125	130	140	150	145	105	130

130	120	102	110	140	98	95
105	140	125	160	133	116	133
135	120					

What is the percentile rank of 158?

SOLUTION: To determine what percentage of the weights are below 158, what do we need to know? How many weights are less than 158, and how many weights are there in total. We can count the total number of weights easily. There are 44 weights. How many of these are less than 158? To determine that we could first place the weights in ascending (or descending) order.

95	98	102	105	105	110	116
120	120	120	125	125	130	130
130	130	131	133	133	135	135
135	140	140	140	140	140	140
145	145	145	147	148	150	155
155	155	155	158	160	160	160
200	200					

From the listing above we see that there are 38 weights which are less than 158. Thus, the proportion of weights less than 158 is

$$\frac{38}{44} = 0.86$$

To express a proportion as a percentage, we simply multiply by 100. Thus the percentile rank of the weight 158 pounds is 86, or the 86th percentile.

EXAMPLE 4.4.7: Using the same data above, determine the weight associated with a percentile ranking of 40.

SOLUTION: The 40th percentile corresponds to a weight which is greater than 40 % of all the weights. Since there are 44 weights, 40 % of 44 is 17.6. What do we do with the fraction? If we take the 17th weight, then there are only 16 weights less than the 17th weight, so that weight really does not satisfy the definition of 40th %tile. [Why?] Therefore, we must "round up" to the 18th weight to find the 40th %tile.

From the list of weights in EXAMPLE 4.4.6, we see that the 18th weight is 133. Thus, the 40th %tile is 133.

Can we divide the data into four equal parts? We would need to break the data so that 25% of the data was in each subset. The first break would be the 25th percentile. [Why?] We know that the 75th percentile would have 25% of the data above it. [Why?] These would be two of our break points. How can we find the third? We need to find the point which will divide the data between the 25th percentile and the 75% percentile exactly in half. That point will be the 50th percentile [Why?] Since these three points divide the data into four equal parts (or quarters), we will call them **QUARTILES**, and denote them by $\mathbf{Q_1}$, $\mathbf{Q_2}$ and $\mathbf{Q_3}$. Q_1 separates the bottom 25% of the data from the top 75%. Q_2 is the median. Q_3 separates the top 25% of the data from the bottom 75%. We formalize this in the following definition:

DEFINITION 4.4.6
The three points which divide a set of data into four equal subsets will be called **QUARTILES**, and denoted by $\mathbf{Q_1}$, $\mathbf{Q_2}$, and $\mathbf{Q_3}$. The bottom 25% of the data will be referred to as the **LOWER QUARTILE**, while the top 25% of the data will be referred to as the **UPPER QUARTILE.**

We can similarly divide data into any number of equal parts. There are many situations in which we can talk about **DECILES** or **QUINTILES**. One extremely important measure of dispersion involves the difference between the upper and lower quartiles, $Q_3 - Q_1$. This gives a measure of how spread out is the middle 50% of the distribution. This measure is known as the **INTERQUARTILE RANGE**. We have the following definition:

DEFINITION 4.4.7
The **INTERQUARTILE RANGE** is the numerical difference between the upper and lower quartile values, $\mathbf{Q_3 - Q_1}$.

SECTION 5 - EXPLORATORY DATA ANALYSIS

The main idea in descriptive statistics is to use simple concepts, diagrams or statistics to "describe" a more complicated set of data. In Chapter 3, we used stem-and-leaf diagrams as a "picture" of a set of data. These pictures gave us a "better feel" of how the data was distributed. In this section we are going to combine some of the statistics we introduced in this chapter into what we call a **5-NUMBER SUMMARY**. This summary describes the data "more fully" than the statistics individually. Building upon the 5-number summary, we will also introduce another pictorial representation of data which attempts to indicate the center of the distribution as well as how spread out it is.

If we were going to select 5 numbers to describe a distribution, what numbers would we choose? To get some idea of the total spread of the data, two candidates would be the **MAXIMUM** ($\mathbf{x_{max}}$), and the **MINIMUM** ($\mathbf{x_{min}}$). A potential problem with these two numbers could be the fact that one or both of these numbers could be so far from the rest of the distribution, so as to not really represent the best picture of the data. One possible solution to that potential problem is to use the **UPPER** and **LOWER QUARTILES**, Q_3 and Q_1. Since the upper and lower quartiles will bracket the middle 50% of a distribution, these numbers will also tell us something about the spread of the middle of the distribution. In addition to the dispersion, we should also be interested in the "center" of the distribution. Two prime candidates to describe the center of a distribution would be the **MEAN**, $\bar{x}$, or the **MEDIAN**, Q_2. Either choice would be appropriate, but in order to provide some sense of consistency we will choose the median. Thus, we have the following convention for the **5-NUMBER SUMMARY**:

CONVENTION 4.5.1

The **5-NUMBER SUMMARY** of a set of data consists of the following numbers:

x_{min}	the smallest value in the data set
Q_1	the lower quartile
Q_2	the median
Q_3	the upper quartile
x_{max}	the largest value in the data set

The following example illustrates how we would determine the 5-number summary of a set of data:

EXAMPLE 4.5.1: Recall the distribution of weights of high school soccer players from EXAMPLES 3.2.6 and 4.4.6. [The weights are given here in ascending order.]:

95	98	102	105	105	110	116
120	120	120	125	125	130	130
130	130	131	133	133	135	135
135	140	140	140	140	140	140
145	145	145	147	148	150	155
155	155	155	158	160	160	160
200	200					

The 5-number summary obtained from the computer is the following:

x_{min} = 95
Q_1 = 125
Q_2 = 137.5
Q_3 = 149
x_{max} = 200

How can we use this 5-number summary? One common way is through a graphic representation known as a **BOX PLOT**. This representation is called a box plot because it uses a box to indicate the data between the lower and upper quartiles. To construct a box plot, first draw a number line scale which will include the minimum and maximum of the distribution. From our example above, we will need to include at least 95 and 200. One possible scale would be the following:

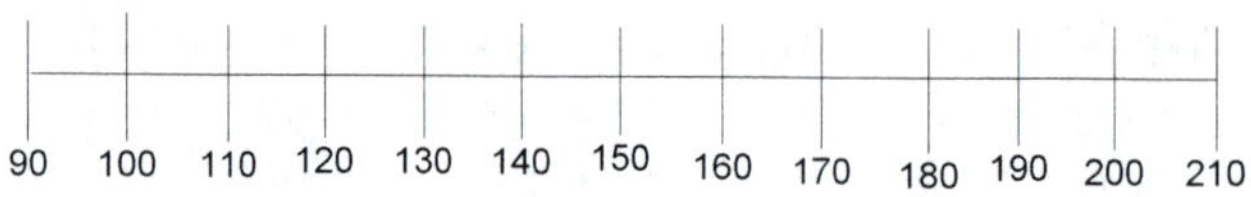

We now locate the lower and upper quartiles on our scale and draw a box with those values as boundaries. It will look like the following:

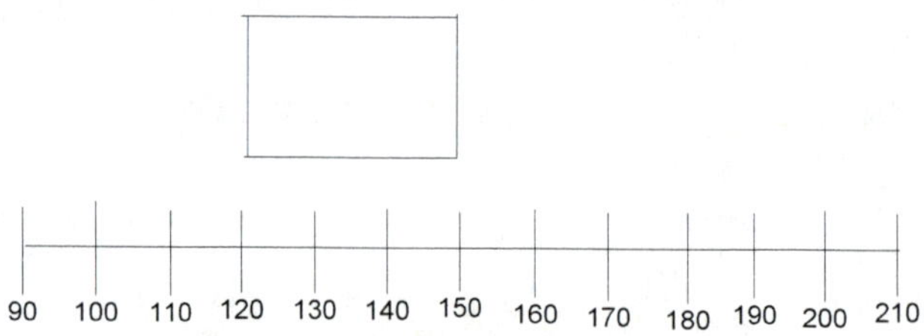

Within the box we now locate the median and draw a line which will divide the box into two parts, as follows:

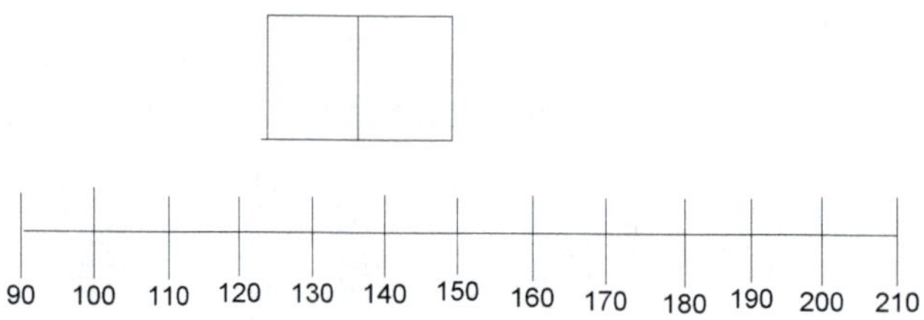

Outside the box we locate the minimum and maximum values and mark with an asterisk (*). We then draw lines from these values to the box as follows:

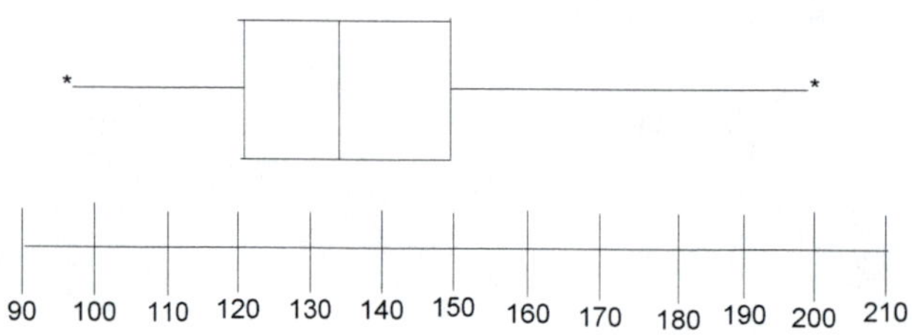

The box represents the difference between the lower and upper quartiles. Earlier in this chapter, we defined that difference as the **INTERQUARTILE RANGE**.

Box plots are useful graphical representations of data. Sometimes they are called "box-and-whiskers" plots, since the lines extending to the maximum and minimum values look like whiskers. Box plots are included in most statistical computer packages. Box plots are relatively simple to construct and yet convey much information about the data. Outliers identified by box plots are not necessarily exactly the same extreme values as those identified by standard scores. However, the simplicity of the box plots make them an easy way to estimate those extreme values which should be investigated.

SECTION 6 - SUMMARY

In this chapter we looked at descriptive statistics. We concentrated on measures of central tendency, dispersion and location. Measures of central tendency are attempts to describe the "center" of a distribution with a single number or value. The three primary measures of central tendency which we considered were the mean, median and mode. The mean is the arithmetic average of all the data points. The median is the point at which half the values are above and half below. The mode is the most populous value. We also looked at the trim mean and the midrange.

Measures of dispersion are attempts to describe the spread of a distribution with a single number or value. The three primary measures of dispersion which we considered were the range, variance and standard deviation. We noted the difference between population and sample variances and standard deviations.

Measures of location are attempts to describe the position of data points within a distribution. We looked at the maximum and minimum values of a distribution. Standard scores and percentiles were also used to describe the location of data within a distribution. Quartiles and the interquartile range were introduced. Finally, 5-number summaries and box plots were used to describe distributions.

KEY TERMS TO KNOW

Measures of Central Tendency
Average
Mean
Median
Mode
Trim Mean
Midrange
Measures of Dispersion
Range
Variance
Standard Deviation
Measures of Location
Maximum
Minimum
Percentile
Standard Score
Quartile
Interquartile Range
5-Number Summary
Box Plot

KEY FORMULAS

TERM	FORMULA
Mean	$\overline{x} = \frac{\sum x}{n}$
Sample Variance	$\sigma^2 = \frac{\sum(x-\overline{x})^2}{n-1}$
Sample Standard Deviation	$\sigma = \sqrt{\frac{\sum(x-\overline{x})^2}{n-1}}$
Population Variance	$\sigma^2 = \frac{\sum(x-\mu)^2}{n}$
Population Standard Deviation	$\sigma = \sqrt{\frac{\sum(x-\mu)^2}{n}}$
Standard Normal Score for Population	$z = \frac{x-\mu}{\sigma}$
Standard Normal Score for Sample	$z = \frac{x-\overline{x}}{s}$

CHAPTER 5
PROBABILITY AND SAMPLING DISTRIBUTIONS

SECTION 1 - INTRODUCTION

In Chapter 3 we introduced the concept of frequency distributions. Recall that a frequency distribution is a listing of the categories or classes of scores and the number or count of how many observations from the population fall into each category or class. In this chapter we will introduce the concepts of probability theory and combine this with the idea of frequency distributions. We will see that probability distributions can be thought of as the theoretical models of what could or ought to occur in the frequency distributions of our sample data. Frequency distributions can be thought of as pictures of *real* or *observed* data, while probability distributions are the pictures of the *possible* outcomes along with their *probable* frequencies. This is laying one more block in the foundation of inferential statistics. At the heart of inferential statistics is the question "How do the actual values compare with the theoretical model?

In Section 2 of this chapter we will introduce the concept of probability. In the next section we will define what we mean by a random variable. We will look at the two major classifications of random variables, discrete and continuous. In Section 4, we will take a closer look at continuous probability distributions. We will discuss the difference between *counting* and *measuring*. In the next section we will look at a specific continuous distribution known as the normal distribution. We will give a formal definition and see what properties of the distribution can be derived from this definition. We will look at a particular normal distribution known as the standard normal distribution. We will see how we can relate any normal distribution to this standard normal distribution.

In Chapter 2, we introduced the concepts of samples and sampling. In many situations, it is not feasible, or even possible, to use the entire population. We must use some subset of the population, i.e., a sample. How do we decide what subset to use? If you recall we discussed several methods of determining what subset to use, i.e., how to sample. The underlying question whenever we sample will be, “How do we know that a sample is sufficient?” or “Why will a sample work?” In this chapter, we will help the student see why one can sample and how large a sample to use.

In Section 6, we introduce the concept of the distribution of means of all possible samples from a given distribution. We also introduce the Central Limit Theorem, by which we can justify the use of samples. In the next two sections, we introduce two types of estimators for means and proportions, point and interval. We also introduce a new family of distributions, known as the t-distributions. Finally, in Section 9 we show, depending upon the situation, how large a sample we need.

SECTION 2 - PROBABILITY

We all use probability everyday of our lives. The meteorologist on the evening news tells us, "There is a 30% chance of rain tonight, and 60% chance of rain tomorrow morning." The sports section of the newspaper reports the odds on each of the final four teams in the 1991 NCAA Basketball Tournament:

UNLV	2:3
UNC	3:1
Duke	8:1
Kansas	20:1

The woman sending in the sweepstakes entry, students discussing the likelihood of getting a date with the homecoming queen, the man purchasing a lottery ticket, the family playing a game of Monopoly, the dorm sign-up lottery, a call for jury duty, and even crossing a busy street are all examples of probability at work.

In this section we attempt to answer the questions, "What is probability?" and "How can we use it?" and we will consider some general and mathematical definitions for probability. The examples with which we began this section had something in common. They all were concerned with the likelihood that a certain event will or will not happen, or did or did not happen. When we talk about the likelihood of something happening, we need to be able to express that likelihood in some manner. We need to be able to measure the likelihood. This leads us to a definition of probability:

> **DEFINITION 5.2.1**
> **PROBABILITY** is a measure of the likelihood that a particular event will occur.

How can we express the likelihood that an event will occur or not occur? On the television game show, Let's Make a Deal, we all knew that the grand prize was behind one of the three doors. We could say that the contestant had a "one out of three chance" of selecting the right door. This is an example of a probability. It is a measure of how likely it was that the contestant would select the right door.

What other ways can we express the same information? The phrase "one out of three" reminds us of fractions. It could be written as the fraction "1/3". We would say that the chance of choosing the right door is 1/3 and the chance of choosing the wrong door is 2/3. Whenever we work with fractions, we can also express them in decimal form, either as a terminating decimal (such as 1/4 = 0.25) or repeating decimal (such as 1/3 = 0.333... or 1/7 = 0.142857142857...). Thus we can express probability as decimal fractions. The probability of selecting the right door is 0.333.... Most of the time when we work with the decimal representation of probabilities, we will round off our probabilities to some predetermined level. If we are rounding off to two decimal places, the probability of selecting the right door is 0.33 and the probability of selecting the wrong door is 0.67.

The use of decimal representations suggests another possible format. The decimal 0.33 can also be expressed as 33%. We can say that the probability of selecting the correct door is 33%. Thus we can express probabilities as fractions (either in rational or decimal format) or as percentages. There is a third format which probabilities can take. Before discussing this format, there are several observation concerning probabilities which we should make.

First, all the probabilities which we have mentioned in this section have had a numerical value between 0 and 1. Second, intuitively, we see that if there is **ABSOLUTELY NO CHANCE** of something occurring, then the probability of that event occurring is 0. If it is **ABSOLUTELY CERTAIN** that something must occur, then intuitively, the probability of that event is 1.

Third, in the Let's Make a Deal example, the probability of the contestant selecting the correct door was 1/3. The probability of selecting the wrong door was 2/3. The sum of these two probabilities is 1. These two events were complementary events. They were the only two events possible in this case. Complementary events are events which are **MUTUALLY EXCLUSIVE**, i.e., events which cannot happen at the same time, and **EXHAUSTIVE**, i.e., events which describe all possibilities. If we let p represent the probability of an event occurring, then the probability of the event's complement occurring is 1 - p. These observations can be formalized into a new definition of probability:

DEFINITION 5.2.2

Given an event A, the **PROBABILITY OF A** is a number P(A) that is assigned to the event A and that has the following properties:

a. $0 \leq P(A) \leq 1$

b. $P(A) = 0$ indicates that the event A cannot possibly occur

c. $P(A) = 1$ indicates the event A occurs with absolute certainty

d. $P(A) + P(\text{not } A) = 1$, were "not A" is the logical opposite or complement of the event A

In the examples of probability which we have used so far, there have been several different types of problems or situations. In the Let's Make a Deal example, there were three doors from which a contestant was to select one. There was no empirical evidence that any one of the three doors or choices was more likely to be correct than the others. This game is exemplary of many games of chance and other similar situations, such as tossing a coin, rolling dice or drawing cards from a deck.

There are several conditions common to all of these games or problems. The trials or "experiments" of the game must be essentially repetitive. This means that we can repeat them under similar conditions. We can toss a coin any number of times, and the potential outcome is not affected by how many times we have tossed the coin before. As long as we replace the drawn card back into the deck and reshuffle the cards properly, the potential outcome of drawing a particular card from a deck of playing cards is not affected by how many times we have already drawn from the same deck.

The second condition which must be met is that we can enumerate every possible outcome of the game or problem. In tossing a coin, there are two possible outcomes, heads or tails. In rolling one die, there are six possible outcomes, one through six. In rolling two identical dice, there are eleven possible outcomes, two through twelve. [If the two die are not identical, but could be distinguished, how many possible outcomes would there be?]

The third condition which must be met is that we can state or determine the expected frequency of each of the possible outcomes of the game. If the coin is "fair", it is equally likely that heads or tails will occur from a toss. If the die is "fair", there is no expected difference among the six possible outcomes. A **one** is as likely as a **two**, or a **six**.

Historically, much of the early work in probability grew out of investigations of games of chance. This type of probability has become known as **CLASSICAL OR THEORETICAL PROBABILITY**. The classical definition of probability is as follows:

> **DEFINITION 5.2.3**
> The **CLASSICAL** or **THEORETICAL PROBABILITY** of an event A, P(A), is given by the quotient of the number of events favorable for event A by the total number of possible outcomes.

This definition can be summarized by the following formula:

$$P(A) = \frac{\text{Number of favorable outcomes to event A}}{\text{Total number of possible outcomes}}$$

In the classical approach to probability, we must assume an ideal world. The coin must be "fair". The die must be "fair". One toss to the next must be consistent. We are also assuming that the structure of the population is known and that we can reasonably assign expected relative frequencies to the outcomes. In such an environment, the probability of an event is interpreted as a theoretical frequency. We do not have to toss a coin to talk about the probability of obtaining a head.

EXAMPLE 5.2.1: What are the probabilities associated with all the possible outcomes of rolling two identical dice?

We first need to identify all the possible outcomes of this experiment. Earlier, we stated that the possible outcomes were the numbers two through twelve. How did we arrive at these outcomes? How can you can get a **two** from rolling two dice? The only way to get a **two** is by having a **one** on each of the two dice. The only way to get a **twelve** is by having a **six** on each die. How can you get a **three**? This question is a littler harder. One way of assuring that the two dice are identical is to roll the same die twice. If we were to do that then we can get a **three** by having a **one** on the first roll and a **two** on the second, or by having a **two** first and a **one** second. Thus there are two ways to obtain a **three**. How can we get a **seven**? To answer that question, we first need to enumerate all the possible outcomes as follows:

DIE 1	DIE 2	TOTAL	DIE 1	DIE 2	TOTAL
1	1	2	4	1	5
1	2	3	4	2	6
1	3	4	4	3	7
1	4	5	4	4	8
1	5	6	4	5	9
1	6	7	4	6	10
2	1	3	5	1	6
2	2	4	5	2	7
2	3	5	5	3	8
2	4	6	5	4	9
2	5	7	5	5	10
2	6	8	5	6	11
3	1	4	6	1	7
3	2	5	6	2	8
3	3	6	6	3	9
3	4	7	6	4	10
3	5	8	6	5	11
3	6	9	6	6	12

In this enumeration how many times does the total **seven** appear? It occurs six times. [What are they?]

What are the probabilities of the eleven outcomes? First we need to count the number of possible ways the one die could be rolled twice. From the table above we see that there are 36 possible combinations. If A represents the event of rolling a total equal to the number A, then the number of favorable outcomes for event A can be determined from the table above. Thus, we can determine P(A) for each A.

$$
\begin{aligned}
P(2) &= 1/36 &&= 0.0278 \\
P(3) &= 2/36 = 1/18 &&= 0.0556 \\
P(4) &= 3/36 = 1/12 &&= 0.0833 \\
P(5) &= 4/36 = 1/9 &&= 0.1111 \\
P(6) &= 5/36 &&= 0.1389 \\
P(7) &= 6/36 = 1/6 &&= 0.1667
\end{aligned}
$$

$$
\begin{aligned}
P(8) &= 5/36 &&= 0.1389 \\
P(9) &= 4/36 = 1/9 &&= 0.1111 \\
P(10) &= 3/36 = 1/12 &&= 0.0833 \\
P(11) &= 2/36 = 1/18 &&= 0.0556 \\
P(12) &= 1/36 &&= 0.0278
\end{aligned}
$$

EXAMPLE 5.2.2: What are the possible outcomes of tossing three coins and what are the associated probabilities?

We first need to identify all the possible outcomes of this problem. One way of displaying the possible outcomes is by using a tree diagram. In tossing the first coin there are two possible outcomes HEADS (H) or TAILS (T). We can illustrate this by the following:

1st Coin

H
T

Now, what happens when we toss the second coin? What are the possibilities?

1st Coin	2nd Coin	Possible Outcomes
H	H	HH
	T	HT
T	H	TH
	T	TT

There are now four possible outcomes. The diagram above illustrates one way to look at the experiment. The outcomes are {(HEAD, HEAD), (HEAD, TAIL), (TAIL, HEAD), (TAIL, TAIL)}. We see that these outcomes could also be described by the number of heads and number of tails. However, the result is not quite the same. If we were to do this, we could have TWO HEADS, ONE HEAD AND ONE TAIL, or TWO TAILS. This appears to be only three outcomes. What happened?

What happens when we include the third coin? We would need to add a third set of branches to the tree diagram. How many outcomes would there be? With one coin, we had two outcomes; with two coins, we had four. To construct the tree diagram we would add two possible outcomes to each of the previously defined outcomes. This has a multiplicative effect. There would be 4 times 2 or 8 branches or outcomes. The tree diagram would be as follows:

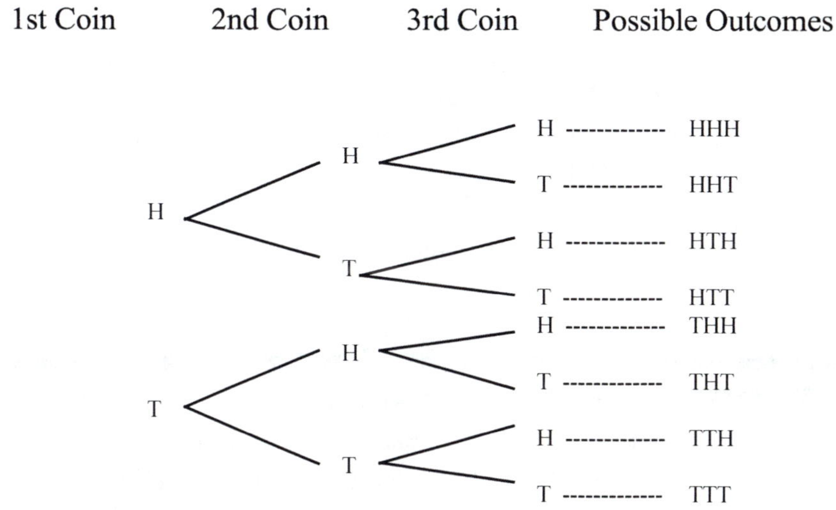

Thus there are eight possible outcomes, if we describe the experiment in this manner. If all three coins are "fair", we would say that the probability of each of these eight outcomes is equally likely, or that the probability is 1/8 or 0.125. If we were to describe the event in terms of the number of heads and tails we would say the outcomes were 3 HEADS, 2 HEADS AND 1 TAIL, 2 TAILS AND 1 HEAD, and 3 TAILS.

What is the probability of getting 2 tails and 1 head? From the tree diagram above we see that there are three ways this situation could occur. The head could be on the first coin, or the second coin, or the third coin. The favorable outcomes are HTT, THT and TTH. Since there are three favorable outcomes out of a possible eight, the probability of getting 2 tails and 1 head is 3/8 or 0.375. [What is the probability of getting 2 heads and 1 tail?]

The multiplicative principle indicated above for situations in which we have repeated trials of an event, such as tossing a coin or rolling a die, is very important. The example of a coin, where there are two and only two mutually exclusive outcomes, can serve as a model for a large number of situations, such as winning or losing game, or the sex of a child. We can call those outcomes SUCCESS and FAILURE. As we increased the number of trials, the sequence of the number of outcomes was 2, 4, 8. What is this pattern? How many possible outcomes are there for four trials?

From the multiplicative principle, we would say that there is 8 (the number of outcomes for three trials) times 2 (the number of outcomes of the fourth trial) or 16 possible outcomes. What is another way to get this number? We repeatedly multiplied 2 to get this answer. How many two's? FOUR! What is another way of expressing the multiplication of two, four times? We can use exponential notation:

$$(2)(2)(2)(2) = 2^4 = 16$$

[How many outcomes would there be for tossing five coins? Ten coins?] Earlier in this section when we were discussing the three doors, we indicated that there was no empirical evidence that any one of the three doors was more likely to be correct than the others. In classical probability, we assume that we know (or can determine) the expected frequency of each of the possible outcomes of the game or experiment. Sometimes, this is not the case.

EXAMPLE 5.2.3: What is the probability that a twenty year-old college student will live to age twenty-one?

The problem can be broken down into a two-outcome experiment: **LIVE**, which we will call a **SUCCESS**, and **DIE**, which we will call **FAILURE**. Intuitively, these two outcomes are not equally likely. How can we determine the relative likelihood of the two outcomes? What happens if we look at a large number of twenty year-old college students, and see how many of them are still alive one year later? Wouldn't this give us an approximate measure of the probability of our given college student making it to his or her twenty-first birthday?

This is the basic approach used by insurance companies in the construction of mortality tables, which are in turn used to help determine life insurance premium rates. For this particular example, suppose the mortality tables say that 9,970 out of 10,000 twenty year-olds will live to be twenty-one. How can we use this information to estimate the likelihood that our student will live? The insurance companies say that there will be 9,970 **SUCCESSES** out of 10,000 trials. Aren't these 9.970 **SUCCESSES** the favorable outcomes which we want? What happens if we apply the classical probability definition to this situation? For this particular example, we would say that the probability of our twenty year-old student living to twenty-one would be:

$$P(A) = \frac{9970}{10000} = 0.997$$

The preceding example leads us to a different type of probability. Our basis of measuring likelihood was the observation of completed experiments or trials. This type of probability has an empirical or experimental basis, and is usually called **EMPIRICAL** or **EXPERIMENTAL PROBABILITY**. The formal definition is as follows:

DEFINITION 5.2.4

The **EMPIRICAL** or **EXPERIMENTAL PROBABILITY** of an event E occurring is found by observing a (large) number of trials of an experiment where E is one of the possible outcomes, and then taking the ratio of the number of times E occurs to the total number of trials

This definition can be expressed by the following formula:

$$P(E) = \frac{\text{Number of occasions E occurs}}{\text{Total number repeats of experiment}}$$

These two definitions are very similar, yet very different. In classical probability, we "know" the probability. In experimental probability, we are trying to determine likelihoods or probabilities. In classical probability we have deduced the likelihood of an event, while in experimental probability we attempt to induce the likelihood of an event. Later in this text, we will see that much of the process of inferential statistics is based on comparing the results of these definitions.

EXAMPLE 5.2.4: How can we determine if a coin is fair? The basic assumption of a fair coin is that the probability of a head is ½ and the probability of a tail is ½. What happens if we toss a coin ten times and get 6 heads and 4 tails. Is the coin fair? Could it be fair?

What happens if we toss the coin ten times and get 9 heads and 1 tail? Would it seems reasonable to think that the coin is fair? What would happen if we tossed a coin 1000 times and got 900 heads? Would that coin appear to be fair?

Going from experimental probability to absolute knowledge is impossible, but the more evidence we have the more reasonable conclusions become. Consider the following story from the prominent 18th century cleric and economist, The Reverend Thomas Bayes:

Let us imagine .... a person just brought forth into this world .... The Sun would, probably, be the first object that would engage his attention: but after losing it the first night, he would be entirely ignorant whether he should ever see it again .... But let him see a second appearance or one return of the sun, and an expectation would be raised in him of a second return, and he might know that there was an odds of 3 to 1 for some probability of this. This odds would increase, as before represented, with the number of returns .... But no finite number of returns would be sufficient to produce absolute or physical certainty.

SECTION 3 - RANDOM VARIABLES

In Chapter 1, we defined a **VARIABLE** as a characteristic or phenomenon which may take on different values and which may be measured or counted in some manner. We also defined an **EXPERIMENT** as the process of selecting a sample and measuring or counting some variable of the sample. With this definition of experiment, sometimes our results are quantitative (such as height, age, counts, test scores); sometimes qualitative (such as gender, ethnicity, political affiliation, socio-economic class). Many times even our qualitative results can be expressed numerically. The gender of an individual can be expressed as **0** if the individual is a female and **1** if the individual is a male. The following example presents a classification system for transforming qualitative data into quantitative data:

EXAMPLE 5.3.1: Our federal government has suggested the use of the following classification system for the ethnic backgrounds of individuals:

ETHNIC GROUP	CLASSIFICATION
Non-Resident Aliens	1
Black Non-Hispanic	2
American Indian/Alaskan	3
Asian or Pacific Islander	4
Hispanic	5
White Non-Hispanic	6
Other	7

We can take all the quantitative problems and those qualitative problems which can be represented or transformed into quantitative ones and make the following definition:

DEFINITION 5.3.1

A RANDOM VARIABLE is a variable that assumes numerical values associated with the outcomes of an experiment.

The following examples are representative of types of random variables:

EXAMPLE 5.3.2: In a family of four children, the number of girls in the family could be: 0, 1, 2, 3 or 4.

EXAMPLE 5.3.3: If we were to toss a coin ten times, the number of tails which could occur would be: 0, 1, 2, 3, 4, 5, 6, 7, 8, 9 or 10.

EXAMPLE 5.3.4: A student's scaled score on the Verbal portion of the SAT exam is any value (whole number value) from 200 to 800, inclusively. In most current situations, SAT scores are only reported to the closest "ten". Therefore, under these conditions, the potential scores are only the "tens" between 200 and 800, i.e., 200, 210, 220,...,790, 800.

EXAMPLE 5.3.5: The weight of an individual (in pounds) could be theoretically any value from 0 pounds and up.

EXAMPLE 5.3.6: The temperature (in degrees Celsius) of water in a glass can be any value between 0°C and 100°C. [Why are the limits 0° and 100°?]

There is a basic similarity among EXAMPLES 5.3.2, 5.3.3 and 5.3.4. There is similarity between EXAMPLES 5.3.5 and 5.3.6. But there is a difference between the two groups of examples. From our work in Chapter 2, we see that the first three are examples of discrete data, while the last two are examples of continuous data. We can make analogous definitions for our random variables.

DEFINITION 5.3.2
A **DISCRETE RANDOM VARIABLE** is a random variable which assumes only a finite or at most countable number of values as the outcome of the experiment.

DEFINITION 5.3.3
A **CONTINUOUS RANDOM VARIABLE** is a random variable which assumes any value in one or more intervals on a line.

A discrete random variable assumes values which are separate, disjoint and can be counted. A continuous random variable can assume infinitely many values, which can be associated with points on a line (continuous scale) in such a way that there are no gaps or breaks. Discrete random variables are **counted**. Continuous random variables are **measured**. Accordingly, EXAMPLES 5.3.2, 5.3.3 and 5.3.4 are examples of discrete random variables. EXAMPLES 5.3.5 and 5.3.6 are continuous random variables.

In the introduction to this chapter we noted that frequency distributions were "lists" of all possible outcomes with the associated number of times each outcome occurred. Probability distributions could then be thought of as "lists" of all possible outcomes with the associated probability of each outcome occurring. How can we "list" all possible outcomes with associated probabilities? The following examples provide us a number of ways of doing this.

EXAMPLE 5.3.7: In a family of four children, the number of possible girls in the family with associated (theoretical) probabilities are given by:

Number of girls	Probability
0	0.0625
1	0.25
2	0.375
3	0.25
4	0.0625

EXAMPLE 5.3.8: In the roll of one die, there are six possible, equally-likely outcomes. The following graph represents the probability distribution for this random variable:

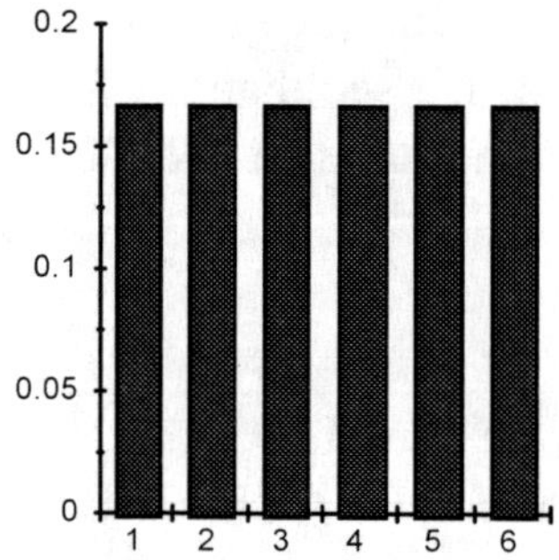

There are several observations which we need to make concerning probability distributions. These observations follow from properties of probabilities which we observed earlier in this chapter. The first observation is a direct carry over from the corresponding property of any probability function: a probability value must be between 0 and 1. If P(x) is a probability function, then

$$0 \leq P(x) \leq 1 \text{ for all values of x}$$

The second observation is a little more subtle. What would happen, if we added all the probability values in EXAMPLE 5.3.7?

$$0.0625 + 0.25 + 0.375 + 0.25 + 0.0625 = 1.0$$

What about EXAMPLE 5.3.8?

$$\frac{1}{6}+\frac{1}{6}+\frac{1}{6}+\frac{1}{6}+\frac{1}{6}+\frac{1}{6}=1$$

Can we generalize from these examples? Recall that the probability of an event which must certainly occur was equal to 1. We can also show that the probability of the union of disjoint events is the sum of the probabilities of the individual events. Thus, if we take all the possible values (disjoint events) and take their union, the sum of those individual probabilities must equal 1 [Why?] Thus, we see that we can use tables, formulas or graphs to represent probability distributions, and that all probability distributions have two properties in common. We can give the following definition:

DEFINITION 5.3.4

A PROBABILITY DISTRIBUTION for a random variable x gives the probability for each possible value of the variable x. It may be given in a table or list, presented as a formula, or represented by a graph or area under a curve. If we let P(x) denote the probability of the value x occurring, then P(x) has the following properties:

a. $0 \le P(x) \le 1$ for all x

b. $\Sigma P(x) = 1$ for all x

SECTION 4 - CONTINUOUS DISTRIBUTIONS

In the previous section, we discussed the concepts of discrete and continuous distributions. In that section, we mentioned that one key difference was that with discrete distributions we **count**, while with continuous distributions, we **measure**. We count the number of heads in the tossing of coins. We count the number of defective items in a sample of manufactured parts. We count the number of correct answers on a True/False quiz. We measure the weight of individuals. We measure the IQ of a group of people. We measure the luminance (sometimes referred to as candle power) of light bulbs.

With discrete distributions, we have discrete objects with which we are working. We can separate them and **count** them. The definition of a continuous variable said that such a variable could assume any value within an interval of values. We can not **separate** things in such a distribution. We try to pinpoint them within the interval. We **measure** them.

Back in Chapter 3 when we were dealing with graphs, we had a similar situation. We noted that certain distributions could be thought of as either discrete or continuous. The one example in particular with which we were dealing was exam grades. If we thought of the grades as letter grades, they were discrete. If we used the numerical grades, we could treat them as discrete or continuous. The grades were discrete if we viewed them as individual integral values. The grades could be thought of as continuous if the numerical grades were thought of real numbers instead of integers. [What is the difference?] The following example (using a previous example) should refresh our memory concerning the difference between discrete and continuous distributions.

EXAMPLE 5.4.1: Recall EXAMPLE 3.4.3.

EXAM GRADE	GRADE INTERVAL	FREQUENCY
A	90 to 100	2
B	80 to below 90	8
C	70 to below 80	9
D	60 to below 70	9
F	Below 60	5

We first illustrated this distribution in the following diagram (Recall DIAGRAM 3.4.1)

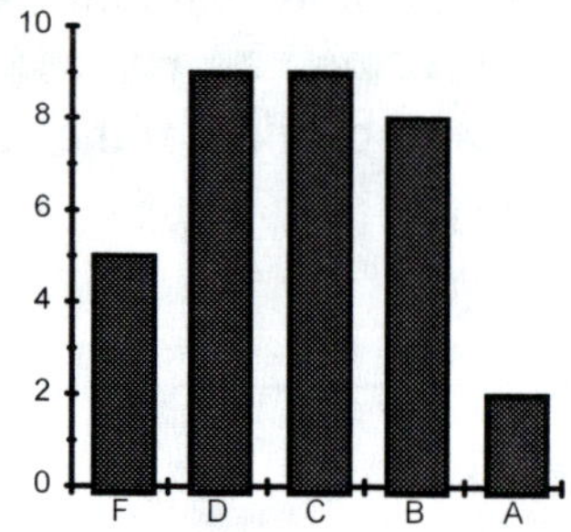

In the diagram above, the bars are disjoint. It would not make any difference "how wide" the bars were. Frequency within a given grade is measured by measuring the height of the bar. In Chapter 3, we went through a discussion of how we could "move the bars" together, use one point (usually the midpoint) as a representative of the bar and construct the following line graph.

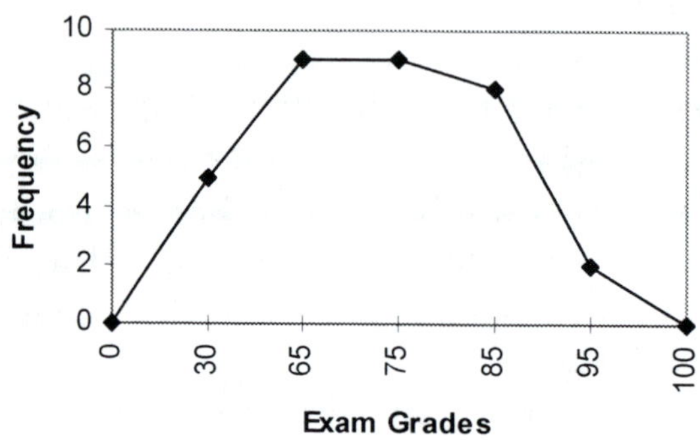

In this diagram, frequency is not really not measured by the height of a line. To find the frequency for an interval of grades is found by measuring the area "under the curve" within the grade interval. The area in any region is the same fraction of the total area as the frequency of that class interval is of the total frequency. In order to do these calculations, we must use geometrical methods or other methods such as the calculus. Although in one sense this makes "the work harder," it will enable us to "do more."

In working with continuous probability distributions, we will find that although there are infinitely many shapes that the distributions may take, there are some general forms which can be easily recognized. The following six diagrams are representative of many probability distributions. Distributions with these shapes occur enough in statistical work, that these shapes are given names.

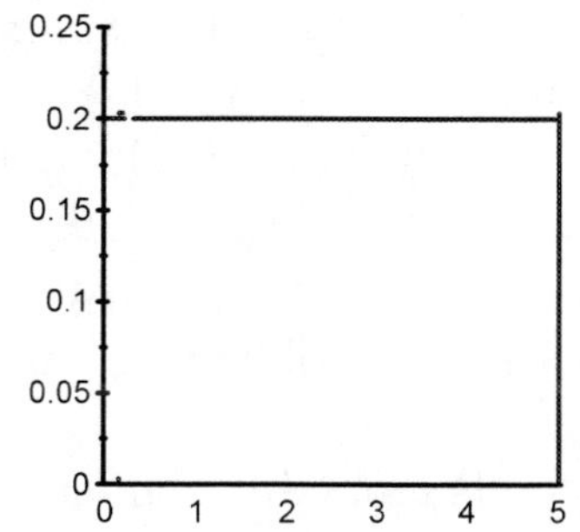

DIAGRAM (a)

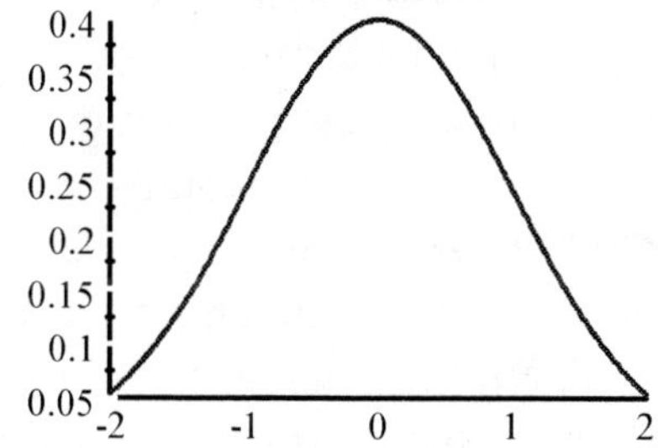

DIAGRAM (b)

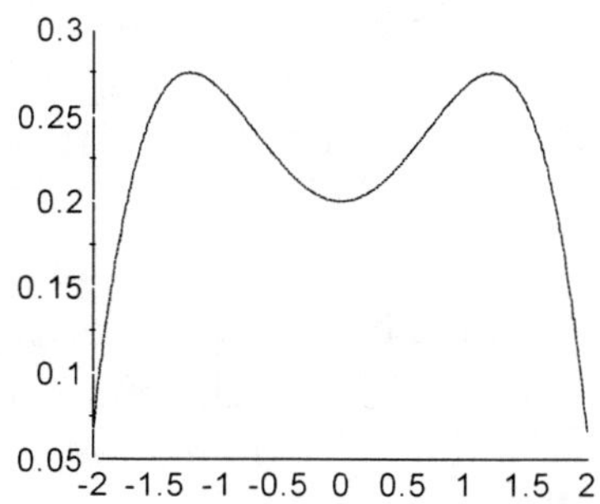

DIAGRAM (c)

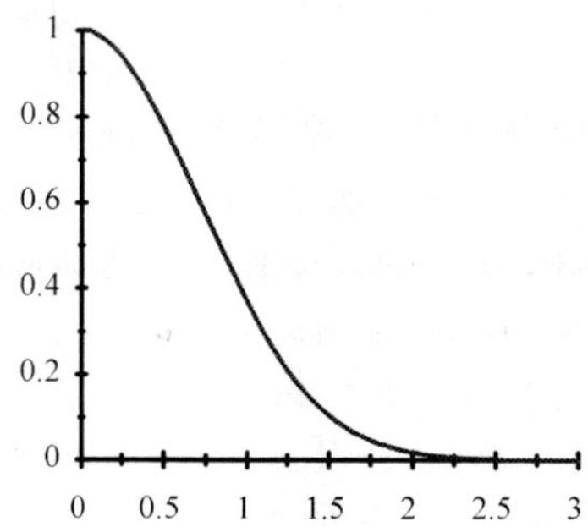

DIAGRAM (d)

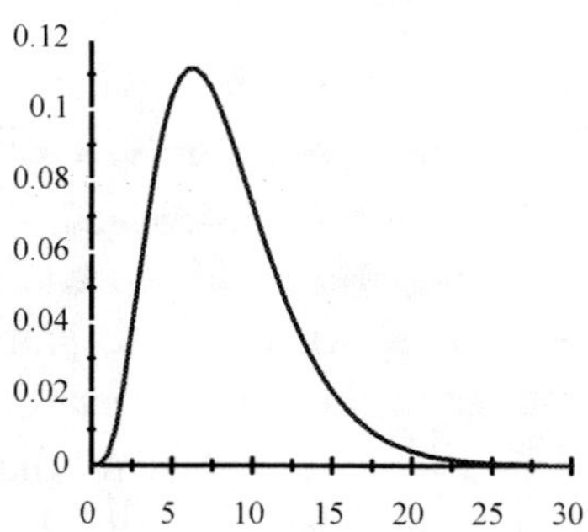

DIAGRAM (e)

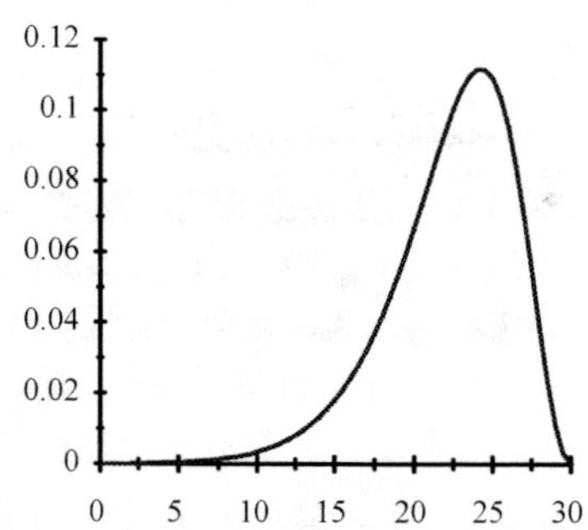

DIAGRAM (f)

The first diagram (a) represents a distribution which is know as a rectangular or **uniform distribution**. In such a distribution all possible values are equally likely to occur. The second diagram (b) is commonly known as a **bell-shaped curve**. [Why?] In such a distribution, value tend to cluster around a central point with central values being much more likely than extreme values. In such distributions, the probability of something occurring above the central point is just as likely as the probability of something occurring below the central point. This particular characteristic is called **symmetry**. In fact, the first two diagrams represent symmetrical distributions. If a distribution is not symmetrical, it is said to be **asymmetric**.

The third diagram (c) represents a distributions which has two points around which values congregate. Such a distribution is called a **bimodal distribution**. Bimodal distributions may be symmetric or asymmetric. The forth diagram (d) represents a distribution in which the majority of values tend to congregate around one of the ends of the distribution, with values trailing off to the other end. Such a distribution is called a **J-curve** or J-shaped distributions, even if the "J" is backwards. J-curves must be asymmetric.

Diagrams (e) and (f) look something like the bell-shaped curve, in that they tend to have a single point around which the majority of the values congregate. The difference occurs in that this point is not in the "center" of the distribution. Thus, "one tail" of the distribution is longer than the other. Such distributions are said to **skewed**. Distribution (e) with the highest point of the graph tending to the left end of the distribution and the right tail longer than the left tail, is said to **positively skewed**. Diagram (f) is **negatively skewed**.

The bell-shaped curves occur everywhere. If we are dealing with large enough populations or samples, bell-shaped curves may be used to describe the weight of adults, the height of adults, IQ scores of people, SAT scores, the life expectancy of light bulbs (how long it takes light bulbs to burn out), or any number of other situations. Since these distributions are so common in nature and life, they have been called **normal curves**. Because of their universality we need to study them more closely.

Shape is just one characteristic of continuous distributions which we need to consider. There are two others characteristics which we considered when we were working with discrete distributions. They were central tendency and dispersion.

When we looked at central tendency, we considered at least three measures of central tendency. The mean was the balancing point of the distribution, the arithmetic average of all the values. We saw in working with discrete probability distributions, that the mean was also the "expected value" of the distribution. The median was the dividing point where 50% of the distribution was above and 50% below. The mode was the most populous point. These same measures can be applied to continuous distributions. In a normal distribution, the three measures are associated with the same point. The highest point of the graph will be the mode. The median and mean are at that same point. In a skewed distribution, the three points will be distinct. In a positively skewed distribution, since the highest point of the graph is on the left side of the graph, the mode will be on the left side. Since the tail of the graph is longer on the right side, the mean and median will be to the right of the mode.

When we looked at dispersion, one of the primary measure we considered was variability. We calculated variance (and hence standard deviation) as a measure of the variability of points in the distribution from the mean of the distribution. In continuous distributions, this type of variability can also be calculated. In the graphs of normal distribution, we will see differences in variability quite easily. The tighter the bell, the less the standard deviation. The fatter the bell, the greater the standard deviation.

SECTION 5 - NORMAL DISTRIBUTIONS

When we introduced the concept of the normal or bell-shaped curve, we described some of the characteristics of it. It is a symmetric curve. The right side of the curve is a mirror image of the left side. The mean, median and mode are at the same place. The center of the distribution is the highest point on the graph. As we go away from the center of the graph, the frequencies decrease. We also noted that differences in the "fatness" of the graph were related to a measure of variability of the distribution.

Given a value of μ and a value of σ, we can draw a normal curve with those characteristics and find probabilities corresponding to any data value. The normal distribution has certain percentages or probabilities between the mean and the mean plus or minus multiples of the standard deviation and has $\mu=0$ and $\sigma=1$. Before computers, data points of the normal distribution had to be first converted to z-scores and then looked up on a table. This will be illustrated conceptually in the next example.

EXAMPLE 5.5.1: The Weschler Adult Intelligence Test (WAIS) is a measure of an individual's (IQ). It is normally distributed with mean $\mu=100$ and $\sigma=15$.

a. Determine the percentage of adults who score between 100 and 120.

If we were doing the problem by hand, we would need to transform the **raw** scores (those scores generated by the intelligence test) into standard scores or **z-scores** using the following formula:

$$z = \frac{x-\mu}{\sigma}$$

The area desired is pictured in the sketch below, with the z-scores on the scale and the raw scores below the scale.

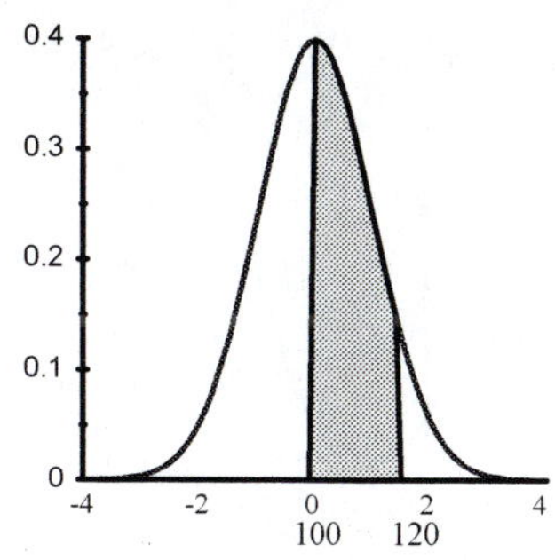

The z-score for x = 100 is 0 and the z-score for 120 is $z = \frac{120-100}{15} = \frac{20}{15} = 1.33$

The area between z = 0 and z = 1.33 can be determined from a table or computer. If the problem is done by hand, the z-score of 1.33 would be used to look up the area (probability) on a table. Some tables give the area between the mean and the z-score; others give the cumulative area. Use the graph to determine what to do with the table or computer value.

If a computer is used, the raw score of 120 can usually be used without converting it first to a z-score. For example, using a raw score of 120, the normal cumulative distribution value would be .9088. To get the answer for the area between 100 and 120, we would subtract .50 from .9088 (Why? .50 or 50% of the area lies to the left of the mean). Therefore the answer is .4088. In other words, 40.88% of adults will have IQ's between 100 and 120.

b. Determine the percentage of adults who score between 90 and 100. The desired area is pictured in the sketch below.

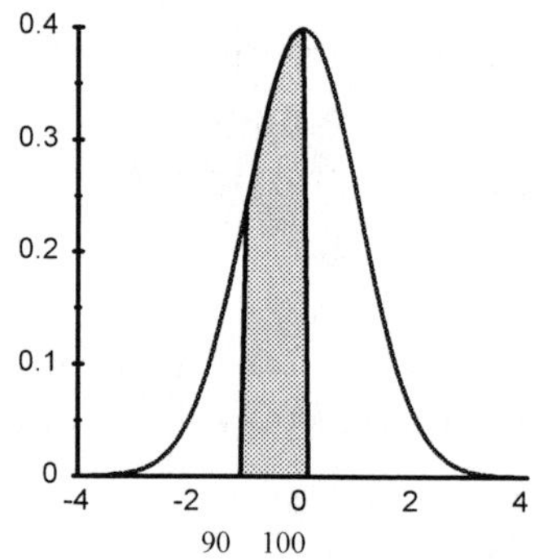

The z-score for 100 is 0. The z-score for 90 is -0.67. There are several ways to proceed:

1) 90 and 100 can be converted to z-scores of -.67 and 0, then the area between them found on a table
2) Using a computer, the cumulative area for 90 is .2525 and the cumulative area for 100 is .50. So the area (probability) between 90 and 100 is .50 - .2525 = .2475. So 24.75% of adults score between 90 and 100.

c. If a score of 130 or higher is considered genius level, what percentage of the population can call themselves geniuses. The area of geniuses is pictured in the sketch below

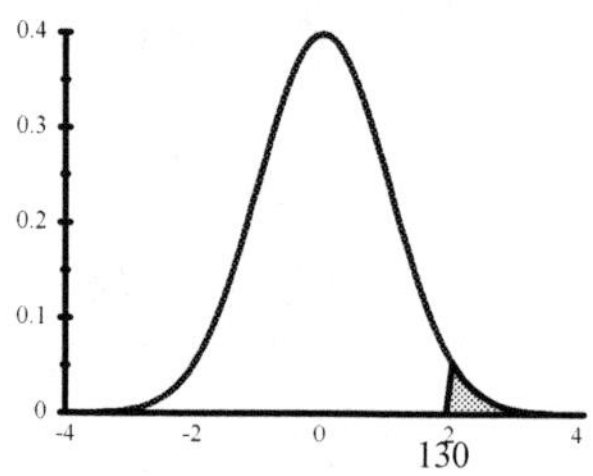

The computed percent for 130 is .9772. Therefore, the percent of the population calling themselves geniuses is 1 - .9772 or .0226 (2.28%) and the area between 0 and 2.0 is 0.4772. Thus, the area beyond 2.0 is 0.0228. Therefore, 2.28% of the adult population can call themselves geniuses.

What happens if we can turn the problem around somewhat and ask the question, "Above what IQ score will the top 25% of adults fall?" The following example will answer this and similar questions.

EXAMPLE 5.5.2: Given the Weschler Adult Intelligence Test (WAIS) with mean 100 and standard deviation 15.

a. Determine the WAIS score above which 33% of adults score. The sketch below pictures what we want to determine.

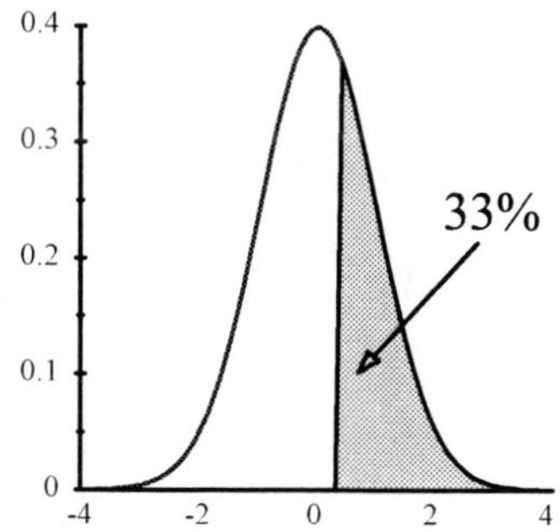

Solution: In previous examples, we had the z-score and found the area. In this problem we have the area and are trying to find the score associated with that area. We want the value below which 67% of the scores fall. This is the same as the 67th percentile.

Doing this problem by hand: Since areas are in the body of a table, we must look there for 0.1700 [Why?]. The z-score associated with an area of 0.1700 is z =0.44. [Why is the value positive?] Since this is a z-score we must translate it to a WAIS or raw score. Taking the z-score formula and solving for x, the raw score, we have

$$x=15z + 100 = 15(0.44) + 100 = 6.6 + 100 = 106.6$$

Thus, the IQ score above 33% of the population score is 106.6.

Doing the problem by computer: 33% in the upper tail is the same as the 67th percentile. Using the computer, we get a score of 106.6.

b. Determine the WAIS score above which 67% of adults score. The following sketch illustrates what we are trying to find.

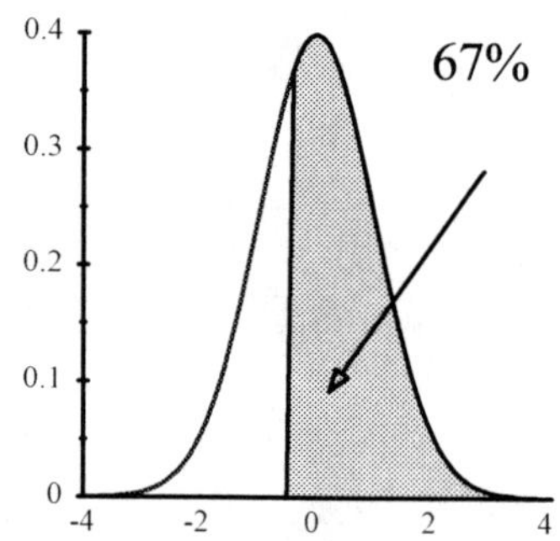

Since 67% is greater than half, the score desired will be below the mean. [Why?] Our value is the same as the 33rd percentile or a probability of .33 on the normal cumulative distribution. This yields 93.4.

SECTION 6 - CENTRAL LIMIT THEOREM

If we were to consider the weight of the soccer players listed in Appendix A.1, we would get a distribution, with mean 137.3 and standard deviation 22.1. What would a graph of the distribution look like? The following is a graph of the individual data points:

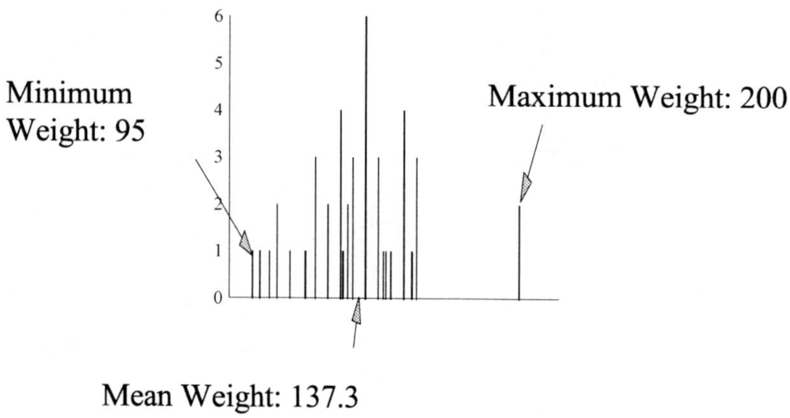

Is there any pattern to this distribution? Does it look like a normal distribution? Even though it does not look much like a normal distribution, can you see the possibility of a "mound-shaped" distribution? What would happen if instead of plotting individual data points, we grouped the data into intervals of length 5? The following diagram plots the distribution for interval lengths of five:

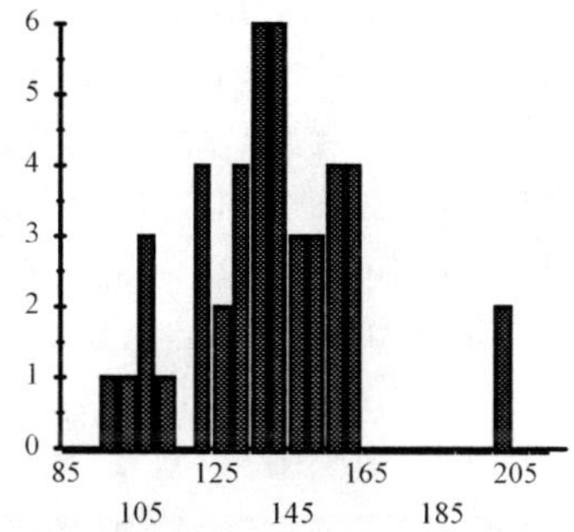

Is there a more discernable pattern? What would happen if we were to group the data in intervals of length 10? What about intervals of length 20? The following graphs illustrate this. Are there any discernable patterns within these distributions? Does one of these graphs "look like" a normal distribution?

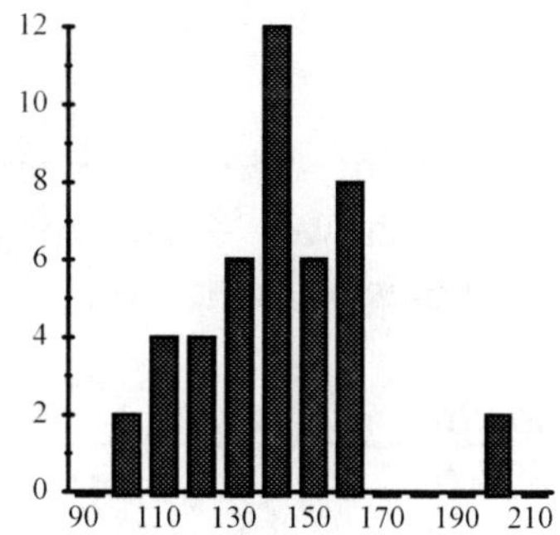

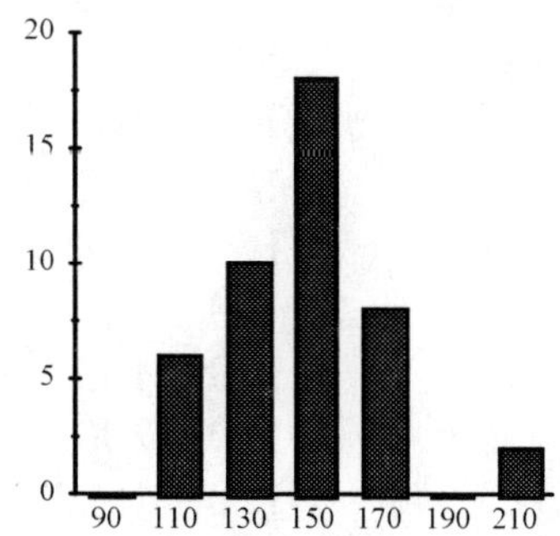

Using intervals of length 20, the distribution of weights of the soccer players "looks" something like a normal distribution. Should this be surprising? What about the distribution of weights of all individuals? Isn't that a normal distribution? Thus we can think of the weights of the players on the soccer team as though they came from a distribution that was "close" to normal, or at least had "some tendencies" toward being a normal distribution.

What would happen if we were to take a random sample of a given size from a normal distribution? From our distribution of the weights of soccer players, one such random sample of size 5 might be: 135, 125, 140, 133, 140. This sample has a mean of 134.6 and a standard deviation of 6.2. Another random sample of size 5 is 130, 200, 160, 158, 131 with a mean of 155.8 and a standard deviation of 28.5. One of these samples has a mean smaller than the population mean, while the other is larger. One of the samples has a standard deviation much smaller than the population standard deviation, while the other is slightly larger than the population standard deviation. Should we be surprised by such a result? What would happen if we took 10 samples of size 5? The following table gives 10 randomly selected samples of size 5 with their means.

1	**2**	**3**	**4**	**5**	**mean**	**1**	**2**	**3**	**4**	**5**	**mean**
135	125	140	133	140	134.6	160	150	110	130	155	141.0
130	158	135	150	155	145.6	130	200	160	158	131	155.8
145	98	155	140	140	135.6	160	148	145	105	133	138.2
135	140	105	140	120	128.0	200	145	200	155	150	170.0
130	130	135	160	160	143.0	125	148	145	200	140	151.6

What happens if we were to find the "mean of the means?" The mean of the ten means is 144.3 with a standard deviation of 12.2. What would happen if we were to take 50 samples of size 5? The following table gives 50 such samples with their means.

1	**2**	**3**	**4**	**5**	**mean**	**1**	**2**	**3**	**4**	**5**	**mean**
135	125	140	133	140	134.6	160	150	110	130	155	141.0
130	158	135	150	155	145.6	130	200	160	158	131	155.8
145	98	155	140	140	135.6	160	148	145	105	133	138.2
135	140	105	140	120	128.0	200	145	200	155	150	170.0
130	130	135	160	160	143.0	125	148	145	200	140	151.6
140	125	200	130	200	159.0	160	160	130	105	155	142.0
95	140	160	148	133	135.2	140	102	120	155	155	134.4
133	150	135	140	147	141.0	155	158	145	140	150	149.6
145	145	135	120	150	139.0	130	102	160	125	150	133.4
148	105	155	131	102	128.2	130	133	158	130	140	138.2
140	120	145	133	155	138.6	120	200	120	155	160	151.0
160	98	130	145	135	133.6	120	150	148	105	155	135.6
131	102	130	133	158	130.8	130	140	140	120	145	135.0
133	155	120	200	120	145.6	120	155	160	160	98	138.6
130	145	135	155	102	133.4	120	155	140	130	155	140.0
140	200	145	148	120	150.6	160	125	120	133	120	131.6
160	95	105	155	131	129.2	160	125	140	98	200	144.6
140	105	160	155	125	137.0	155	105	155	110	140	133.0
160	145	155	130	130	144.0	148	120	158	145	125	139.2
145	140	155	200	131	154.2	120	130	130	200	145	145.0

130	140	145	125	105	129.0	155	95	120	160	133	132.6
95	120	160	133	160	133.6	133	95	120	160	155	132.6
120	160	160	130	95	133.0	133	200	155	130	120	147.6
200	110	160	155	105	146.0	105	98	110	140	135	117.6
131	102	120	155	145	130.6	125	148	148	116	145	136.4

They are not all 137.3. [In fact, not one of them is 137.3.] However, they are for the most part, close to 137.3. In fact, they seem to congregate "somewhere around" 137.3. The actual mean of the 50 means is 139.4, which is close to 137.3, and the standard deviation of the 50 values is 9.28, which is much smaller than 22.1. What would happen if we plotted these 50 sample means? Using an interval length of 5, the following diagram illustrates such a graph.

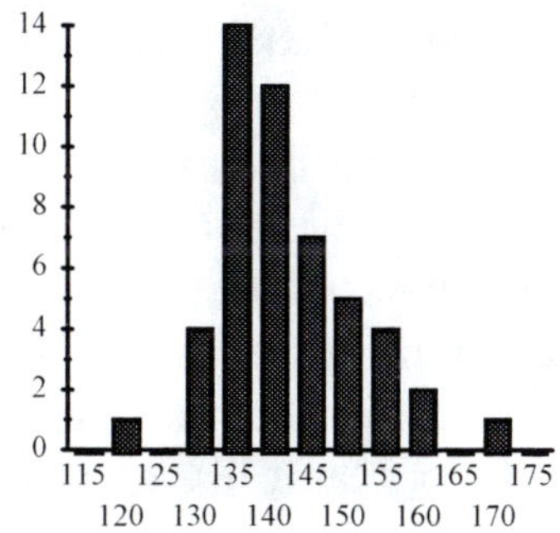

What does this graph suggest? What is the general shape of the distribution? Does it look like a normal distribution? What is the relationship between the population mean and the mean of the sample means? Is the population mean greater than, less than or equal to the mean of the sample means? What is the relationship between the standard deviation of the population with the standard deviation of the distribution of sample means? Is the distribution of sample means "tighter", "more spread out" or the same as the original population?

Can we suggest that the distribution of sample means will fall into a normal distribution with mean close to or equal to the mean of the original distribution, but with a standard deviation which is smaller than the original distribution? Can this result be generalized? The answer to both questions is **YES!** The generalization of this result is one of the main theorems or building blocks of statistics. It is known as the **CENTRAL LIMIT THEOREM**.

THEOREM 5.6.1 - CENTRAL LIMIT THEOREM

Given random samples of size n, drawn from a population with mean μ and standard deviation σ, the resulting distribution of sample means, $\bar{x}$, tends to be normally distributed with mean μ and standard deviation $\frac{\sigma}{\sqrt{n}}$.

If we let $\mu_{\bar{x}}$ and $\sigma_{\bar{x}}$ represent the mean and standard deviation of the distribution of sample means, then the Central Limit Theorem provides us with the following formulas:

$$\mu_{\bar{x}} = \mu \qquad \text{and} \qquad \sigma_{\bar{x}} = \frac{\sigma}{\sqrt{n}}$$

We introduced the statement of the theorem to help illustrate the direction we want to go. If the underlying population is normal, the resultant distribution of means will be normal. If the underlying population is not normal, then the resultant distribution of means will be approximately normal. The larger the sample size, the closer to normal it will be. Thus, when we look at problems, we need to ask ourselves several questions which will help us to be able to apply the Central Limit Theorem correctly.

First, **IS THE UNDERLYING DISTRIBUTION NORMAL?** If the distribution from which we drawn a sample or samples is normal, or essentially normal, then the distribution of sample means will be normal, or essentially normal, and we can use the full power of the Central Limit Theorem. We will be able to estimate the mean of the distribution of sample means by the mean of the original distribution. Also, the standard deviation of the distribution of sample means can be estimated by the standard deviation of the original distribution divided by the square root of our sample size. For most purposes, if a distribution is almost normal, i.e., if it "looks" like a normal distribution, then we will be able to treat it like a normal distribution. Any approximations which we make will be "close enough."

If, the underlying distribution is not normal, what happens? The Central Limit Theorem can still help us, if the sample size is large enough. The second primary question is **WHAT SAMPLE SIZE IS LARGE ENOUGH**? There is no one uniform (or easy) answer to that question. The larger the sample, the better the estimates generated by the Central Limit Theorem will be. Although, the effectiveness of the approximations vary from example to example, we can use the following guideline or convention:

CONVENTION 5.6.1

In applications of the Central Limit Theorem, no matter what the underlying distribution is, if the sample size is greater or equal to 30, i.e. **n≥30**, then the distribution of sample means is sufficiently close to normal to assume that it is normal, with mean μ and standard deviation $\sigma / \sqrt{n}$.

The following example illustrate how the sampling distributions of for two populations begin to look very much like normal distributions, even though the original distributions are very different from normal distributions.

EXAMPLE 5.6.1: Sampling Distributions of $\bar{x}$ for Different Populations and Sample Sizes

Original Populations

I **II**

Sampling Distributions of $\bar{x}$ for n = 2

Sampling Distributions of $\bar{x}$ for n = 5

Sampling Distributions of $\bar{x}$ for n = 30

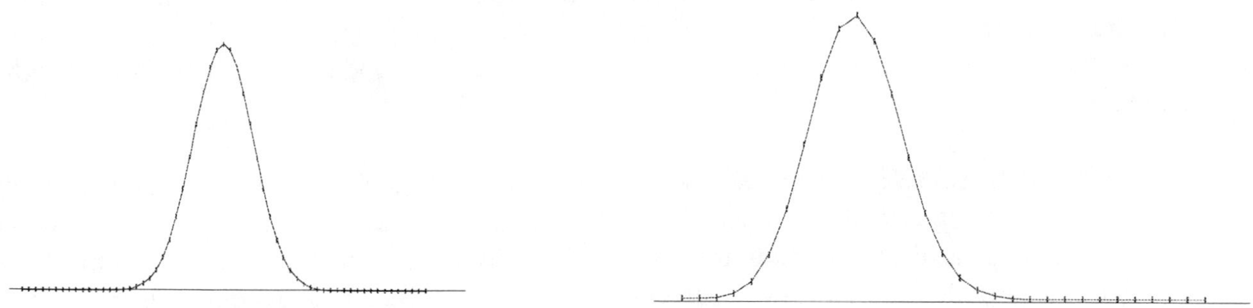

Why is the Central Limit Theorem important? We have already noted that the Central Limit Theorem deals with three aspects of a distribution: the center (mean), the variability (standard deviation) and the shape. Do we have to be working with normal distributions to use the Central Theorem? From the two examples above, we can see that even though the shape of our original distributions were very different, the distributions of sample means became closer and closer to normal. We will make use of this very important result to infer information concerning population parameters without KNOWING the shape of the distribution. In "the real world," we almost never know the exact shape of the distribution with which we are working.

Suppose we are working with a distribution for which we do not have complete information. We take a large sample from this distribution, and calculate the mean and standard deviation of the sample. A logical question is: "How close do we think the sample mean is to the actual mean?" To answer this question, we will make use of the Central Limit Theorem. The Central Limit Theorem says that the standard deviation of the distribution of sample means is $\sigma / \sqrt{n}$. This quantity will be called the **STANDARD ERROR OF THE SAMPLE MEANS**, or just **STANDARD ERROR**, and denoted by $\sigma_{\bar{X}}$. We have the following definition:

DEFINITION 5.6.1
The **STANDARD ERROR OF THE MEAN,** $\sigma_{\bar{X}}$ also known as the **STANDARD ERROR**, is the standard deviation of the distribution of the means of all samples of a given size n.

As long as we know the standard deviation of the original population, **EQUATION 5.6.2** provides us with the computational formula for the standard error. Why is this thing called a **STANDARD ERROR**? We started with a given population. We took a number of samples of a fixed size n. We then calculated the mean of each sample. These means are estimates (statistics) of the mean (parameter) of the entire population of means of samples of size n, drawn from the given population. The standard deviation of any distribution can be thought of the average amount by which the data points deviate or vary from their mean. The statisticsx are estimating the parameter μ. Hence the deviations of from μ are in a real sense errors. Thus the standard error represents a measure of deviation of the sample means from the population mean. It is a measure of the variability of the estimate .

The following example shows us how to calculate the standard error and how it might be used in a specific situation:

EXAMPLE 5.6.2: Suppose that in fifteen years of teaching statistics, an instructor has attempted to give examinations which were of comparable difficulty. Using all classes and all students as the population, the instructor has found that the mean score was 62.4% with a standard deviation of 14.7 percentage points. In the most recent January term, the instructor felt that a class of 37 students was a "really good" class. This particular class had a mean of 69.2%. How does this class compare with the population of all students from this instructor's classes?

Now the population mean is $\mu = 62.4$, the population standard deviation is $\sigma = 14.7$, and the sample mean is $\bar{x} = 69.2$. From our computational formula, the standard error for the distribution of sample means of size 37 [Why 37?] is

$$\sigma_{\bar{x}} = \frac{14.7}{\sqrt{37} = 2.42}$$

Therefore, we can say that the sample mean (for this class) is

$$z = \frac{\bar{x} - \mu}{\sigma_{\bar{x}}} = \frac{69.2 - 62.4}{2.42} = 2.81$$

standard errors away from the population mean. How often would that occur by chance? From our work in the previous chapter, we know that the area under the normal curve above a z-score of 2.81 is only 0.00248. What does that mean? Only 2.48% of the time will a class achieve a class mean above the given class mean. Thus, it is not very likely that this class was a "typical" class. The instructor was right. It was a "really good" class.

As we mentioned above, the term **STANDARD ERROR** carries with it the idea of deviation from the population mean. We will be using sample means to estimate a population mean. Do they always agree? Do they ever agree? In reality, they will almost never agree. The difference between a sample mean and the population mean can be attributed to at least two distinct types of **ERRORS**. One type of error may be built into the sample selection process. Back in Chapter 2, we talked about systemic or systematic errors in the selection of samples. If we can identify the source of systematic error, then we can work to eliminate that error. On the other hand, some errors may occur no matter what we do. These are random in nature.

If we look closely, both types of errors are included in the statement of the Central Limit Theorem. We will refer to the two types of error as **ACCURACY** and **PRECISION**, and define them as follows:

> **DEFINITION 5.6.2**
> **ACCURACY** is the absence of systematic error.
> **PRECISION** is the absence of random error.

Where is accuracy included in the Central Limit Theorem? Since $\mu = \mu$, the sample mean is an unbiased estimate of the population mean. By definition it is therefore free from systematic error. Thus, the Central Limit Theorem tells us that using sample means as estimates of a population mean is an accurate process.

Where is precision included in the Central Limit Theorem? Does the Central Limit Theorem tells us anything about the random errors which occur? The Central Limit Theorem does indicate that the standard error is given by σ/√. Since we are dividing the population standard deviation by a number which is always greater than or equal to 1 [Why?], the standard error will never be larger than the standard deviation of the population. The variability of the distribution of sample means is less than the variability of the original population.

The standard error is a measure of the precision [the variability or freedom from random error] of the sample mean as an estimate of the population mean. With the size of the sample in the denominator of the expression for the standard error, we note that as the sample size increases, the standard error decreases. As the sample size increases, the sample mean becomes a more precise estimate of the population mean.

We need to make one very important note at this point. By saying that the sample mean becomes more precise, we are not saying that, in a given situation, one sample mean will necessarily be "very close" to the mean of the population from which the sample was drawn. What we are saying is that, by increasing the sample size, we are "more likely" to have a better estimate of the population mean.

So far in our discussions about the Central Limit Theorem, there is one aspect concerning the relative size of our population and our samples that we have not mentioned at all. In the statement of the Central Limit Theorem, we used the sample size n but made no mention of the population size. Statisticians make an implicit assumption about the size of the population. They assume that it is very large in relationship to the samples being drawn. If we are drawing samples with replacement, then this is a reasonable assumption. If we replace units before drawing the next sample, even before drawing the next element of a given sample, the population will never "run out."

SECTION 7 - POINT ESTIMATES

Estimation is the process of using a value or values to approximate a population parameter. How do we know what values we can or should use to approximate a given parameter? Having selected a particular value, how do we know how "good" an estimator it is? The simplest form of estimation is to use a single value to estimate a population parameter. We have the following definition:

> **DEFINITION 5.7.1**
> A **POINT ESTIMATE** is a single numerical value which approximates or is taken to be indicative of a population parameter.

The definition of point estimate tells us that we are going to use a single number to approximate a parameter. It does not tells us how to determine the point estimate. The examples below may provide us some direction in this problem:

EXAMPLE 5.7.1: For all 32-ounce dish detergent bottles, a point estimate of the mean of net contents could be 32 ounces. This point estimate is determined by outside information. We are "trusting" the manufacturer to have told us the "truth" in saying that the bottles contained 32 ounces of detergent.

EXAMPLE 5.7.2: For the detergent in EXAMPLE 5.7.1, suppose we take one of those 32-ounce bottles. We measure the net contents for this bottle, and find it to be $\bar{x} = 32.3$ ounces. [Why can we use the notation for sample mean for the net contents of the one bottle?] The value $\bar{x} = 32.3$ can be used as a point estimate for the population parameter of mean contents.

EXAMPLE 5.7.3: For the detergent in EXAMPLE 5.7.1, suppose we take a sample of 100 of those 32-ounce bottles. We calculate the mean net contents for these 100 bottles, and find it to be $\bar{x} = 32.1$ ounces. The value $\bar{x} = 32.1$ can be used a point estimate for the population parameter of mean contents.

In two of the three examples above we used sample means as point estimators for the population mean. Why might this be appropriate? The Central Limit Theorem actually encourages us to do this. The Central Limit Theorem tells us that the mean of the distribution of the means of all samples of a given size equals the mean of the distribution, i.e., $\mu = \mu_{\bar{x}}$.

What is the problem with this? It would be very unusual if we know the mean of the distribution of sample means. Most of the time, we will have only the one sample mean, $\bar{x}$, with which we are working. What do we do? What happens if we use $\bar{x}$ to estimate μ, which the Central Limit Theorem says is an estimate of μ? In effect, we are using $\bar{x}$ as a point estimate of μ.

How good an estimator of μ is $\bar{x}$? In the previous section we discussed the two concepts of accuracy and precision. The question, "How good an estimator of μ is $\bar{x}$?" can be translated into two questions: "How accurate an estimator of μ is $\bar{x}$?" and "How precise an estimator of μ is $\bar{x}$?" The question of accuracy depends upon the nature of the sample and how the data was collected. If the sample under consideration was randomly selected and the sample statistic unbiased, then we would have as accurate an estimator as we could have. i.e., as free of systematic error as possible. Does this mean that $\bar{x}$ = μ? Does it even mean that the estimate is free from random error? No, it means that any difference between the estimate and the parameter is due to random error and not due to something inherent to the sample or process.

But what about this random error? This is the concern of the question "How precise an estimator of μ is $\bar{x}$?" Recall that the standard error

$$\sigma_{\bar{x}} = \frac{\sigma}{\sqrt{n}}$$

is a direct measure of the precision of the means of random samples of size n. It represents the "average" amount by the mean of a sample of size n differs from the population mean. What is the problem with this formula? Most of the time we do not know the population standard deviation σ. Most of the time, we will only know the standard deviation s of one sample. Thus, the best we can do is to use the **ESTIMATED STANDARD ERROR OF THE MEAN**,

$$s_{\bar{x}} = \frac{s}{\sqrt{n}}$$

This tells us something about the precision of the estimate. The smaller the standard error (or estimated standard error), the more precise the sample mean is as an estimator of the population mean. To see how we might use accuracy and precision to provide additional information about our point estimates, consider the sample of 100 detergent bottles in EXAMPLE 5.7.3:

EXAMPLE 5.7.4: Suppose the sample of 100 bottles of detergent was selected in the following manner: 100 cases (of 24 bottles each) of the detergent were randomly selected from cases ready to be shipped. From each case one bottle was randomly selected and its contents measured. The sample mean for this sample was found to be $\bar{x}$ = 32.1 ounces with a sample standard deviation of s = 0.2 ounces. How accurate and precise is this sample mean as an estimator for the population mean?

Since we have selected the 100 bottles in a random manner, and any bottle ready to be shipped, had an equal opportunity to be selected, we have a random sample. The sample mean should be free of systematic error. It will be as accurate an estimator of the population mean as we can find.

The sample standard deviation can provide one measure of the precision of the estimate. Since s = 0.2, we can calculate the estimated standard error as follows:

$$s_{\bar{x}} = \frac{s}{\sqrt{n}} = \frac{0.2}{\sqrt{100}} = \frac{0.2}{10} = 0.02$$

This would seem to indicate that the sample mean is "fairly precise" as an estimator of the population mean.

What is the problem with the last statement above? Although the sample mean is "fairly precise" as an estimator of the population mean, there will still almost always be some difference between the two. The use of a sample mean introduces an error. The single number, however, gives the impression of accuracy and precision. We need to be always aware that **ESTIMATES** are just that -- **ESTIMATES**. When we use point estimates we should indicate some information about the accuracy and precision of our estimates.

How can we present point estimates and at the same time indicate information about the precision of that estimate? There are two common ways of doing this. One way is to present the point estimate with the estimated standard error as follows:

sample mean ± estimated standard error of the mean

In the example above, we could have reported our results as: The sample mean was 32.1, with an estimated standard error of ± 0.02.

If we are using a sample mean as our point estimate, a second way of presenting information about the precision of the estimate is to also report the sample standard deviation, i.e.,

sample mean is $\bar{x}$ with sample standard deviation s

In the example above, we could have reported our results as: The sample mean was 32.1, with a sample standard deviation of 0.2.

Since point estimates have inherent or built-in errors, is there another way of estimating population parameters? The two examples above suggest the possibility of using intervals. The next section indicates how we can construct such intervals.

SECTION 8 - INTERVAL ESTIMATES

In the previous section, we introduced point estimates for population parameters. We showed how these estimates have inherent imprecision. Even when additional information is presented which indicates how imprecise the estimate might be, the reader is left to interpret the results. What we need is some measure of the difference between the population parameter and the point estimate which we are using. How can we get at such a measure? One way is through the use of the following definition:

> **DEFINITION 5.8.1**
> The **SAMPLING ERROR** is the magnitude of the difference between an unbiased point estimator and the corresponding population parameter.

When we use a sample mean to estimate a population mean, we have the following:

$$\text{Sampling error} = |\bar{x} - \mu|$$

What does $|\bar{x} - \mu|$ mean? The lines indicate ABSOLUTE VALUE. The absolute value of a given value is the "positive" distance from the origin or the magnitude of the value, i.e. $|3| = |-3| = 3$. How can this help when most of the time we will not know what the population mean μ is? The Central Limit Theorem tells us that for large samples the sampling distribution of sample means is approximately normal with mean $\mu_{\bar{x}} = \mu$ and standard deviation of $\sigma_{\bar{x}} = \dfrac{\sigma}{\sqrt{n}}$.

So what? Even though we do not know what the population mean μ is, we do know how the values of the sample means are distributed about the unknown μ. For example, we know that 95% of all values in a normal distribution are within ± 1.96 standard deviations of the mean. Therefore, if we are looking at the distribution of all sample means $\bar{x}$ (for a given sample size n), then we know that 95% of all $\bar{x}$ values are within $\pm 1.96\sigma_{\bar{x}}$ of the population mean μ. We can illustrate this in the following diagram:

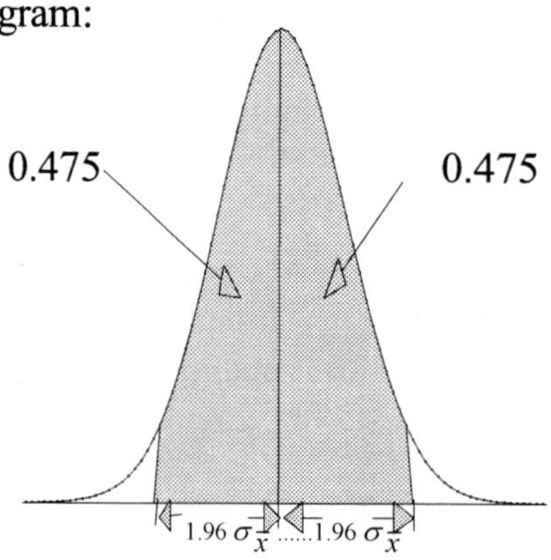

In effect the interval $\pm 1.96\sigma_{\bar{x}}$ provides a measure of precision of our estimate of the population mean. We can illustrate this in the following example:

EXAMPLE 5.8.1: Suppose as a consumer testing service we are interested in the life expectancy of 60-watt light bulbs. We select a sample of 100 light bulbs from a particular manufacturer. We measure how long it takes for each of the 100 light bulbs to burn out. For this sample of 100 light bulbs, we find a sample mean of = 1038.2 hours.

From the discussion above we know that 95% of all sample means from samples of size n = 100 will be within $\pm 1.96\sigma_{\bar{x}}$ of the population mean. To determine this interval, we need to know $\sigma_{\bar{x}}$. The manufacturer states on the package of light bulbs that the "average life expectancy" of its light bulbs is 1000 hours with a standard deviation of 250 hours. If we assume that the population standard deviation σ is indeed 250 hours, then we have the following:

$$\sigma_{\bar{x}} = \frac{\sigma}{\sqrt{n}} = \frac{250}{\sqrt{100}} = 25$$

Substituting this value into our interval $\pm 1.96\sigma_{\bar{x}}$, we can assume that 95% of all sample means of size n = 100 will be within ± 1.96(25) = ± 49 hours of the population mean μ. This means that there is 0.95 probability that a sample of 100 light bulbs will provide a sampling error of 49 hours or less.

The sampling error really represents a measure of the precision of the estimate. Is there a general procedure for making probability statements about the sampling error whenever we use $\bar{x}$ as an estimate of μ? The preceding example suggests such a procedure. Let α indicate the probability that a sampling error is greater than a specified value, and let $z_{\alpha/2}$ represent the standard normal score at which the area under each tail of the distribution is α/2. From the following diagram, we see that 1 - α is the probability (area under the curve) that a sample mean will give us a sampling error less than or equal to the specified sampling error, and α/2 is the probability (area under the curve) in each tail of the distribution.

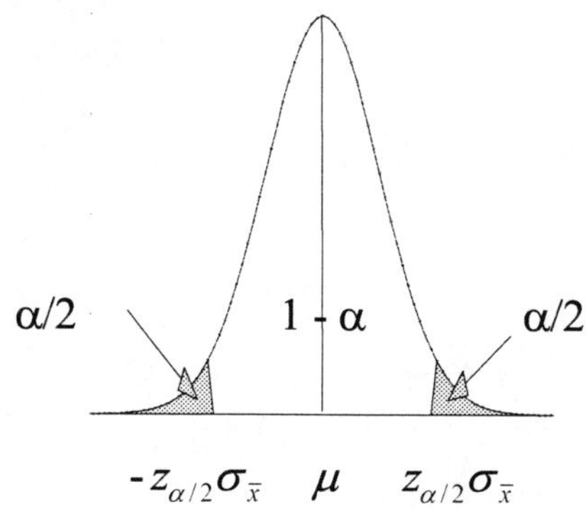

In this case we can say that whenever a sample mean $\bar{x}$ is used to estimate the population mean μ, there is a 1 - α probability that the value of a sample mean will provide a sampling error of $z_{\alpha/2}\sigma_{\bar{x}}$ or less. What do we know about the interval

$$(-z_{\alpha/2}\sigma_{\bar{x}}, z_{\alpha/2}\sigma_{\bar{x}})?$$

It is an interval which we believe contains the population mean μ. It is a type of ESTIMATE which we will call an INTERVAL ESTIMATE. An interval estimate can be defined as follows:

DEFINITION 5.8.2
An **INTERVAL ESTIMATE** of a population parameter is an interval which we assert contains the population parameter under consideration.

Since we usually do not know the population mean, what happens if we use a sample mean to determine a point estimate of the population mean? From the discussion above we know that for the particular population at which we are looking, we have a sampling error of $z_{\alpha/2}\sigma_{\bar{x}}$ or less. What does this mean? If we were to take repeated samples of the same size from this population and construct all intervals centered at the corresponding sample mean with a length equal to two times the sampling error, i.e., construct all intervals of the form

$$(\bar{x} - z_{\alpha/2}\sigma_{\bar{x}},\ \bar{x} + z_{\alpha/2}\sigma_{\bar{x}})$$

then 100(1 - α)% of these intervals will contain the population mean.

Consider the light bulb problem posed in **EXAMPLE 5.8.1**. What happens if we take many different samples of size 100 and construct the corresponding intervals, i.e., intervals which are centered at the corresponding sample mean and which are of length (2)(49) = 98 hours? How many of these intervals will contain the population mean and how many will not?. Our previous discussion tells us that 95% of all intervals which we can construct in this manner will contain the population mean. But if 95% of the intervals contain the population mean then 5% of all such intervals will not. We can say that we have 95% **confidence** that such an interval estimate will contain the population mean. The interval is referred to as a **95% confidence interval** and the value 95% is the **confidence coefficient**. We can generalize these statements into the following definition:

DEFINITION 5.8.3

Suppose samples of size n are drawn from a population with mean μ and standard deviation σ. Given a value α, the interval

$$(\bar{x} - z_{\alpha/2}\sigma_{\bar{x}}, \bar{x} + z_{\alpha/2}\sigma_{\bar{x}})$$

is known as the **(1-α) CONFIDENCE INTERVAL** estimate of the population mean. The value 1-α is called the **CONFIDENCE COEFFICIENT.** The confidence coefficient expressed as a percentage is known as the **CONFIDENCE LEVEL.**

What is a confidence interval? When we say that we have constructed a 95% confidence interval, we are not really implying that there is a 95% probability that the given interval contains the population mean. We are making a statement about how the interval was constructed. We are saying that 95% of all intervals which can be constructed in the prescribed manner, will contain the population mean. We are confident in the process, not in a given interval. However, if we have a certain confidence in the process, can't we *transfer* some of that confidence to a given interval which was the result of that process? Thus, when we have constructed a particular confidence interval we can (and usually do) act as though we are saying the following: "We are 100(1 - α)%

confident that the particular interval contains the population mean." In later chapters we will be making decisions based on this assumption.

Given a particular population and a particular sample, what is the relationship between the confidence level and the size of the confidence interval? To look at this aspect of confidence intervals, let us reconsider to the light bulbs of **EXAMPLE 5.8.1**:

> **EXAMPLE 5.8.2:** Suppose we look at the life expectancy claim of a light bulb manufacturer. This particular manufacturer claims that the "average" life expectancy of its 60-watt bulbs is 1000 hours with a standard deviation of 250 hours. Suppose we take a sample of 100 bulbs and find the sample mean of this particular sample to be $\overline{X}$ = 1038.2. What would happen if we constructed a 95% confidence interval for this sample?
>
> By hand, from our previous discussions we know that the 95% confidence interval for this sample would be
>
> $$(\overline{x} - z_{\alpha/2}\sigma_{\overline{x}}, \overline{x} + z_{\alpha/2}\sigma_{\overline{x}})$$
>
> For this particular example, this gives us a confidence interval of $1038.2 \pm 1/96(250/\sqrt{100})$, yielding the interval of 989.2 to 1087.2.
>
> By computer, select a confidence level of 95% and the answer will be the same as above.

What would happen if we select a confidence level of 99%? We would get (973.7, 1102.7). Comparing the two intervals, we see that the 99% confidence interval is larger than the 95% confidence interval. By increasing the confidence level, we make the confidence interval larger.

What would happen if we select a confidence level of 90%? Our confidence interval would be (996.95, 1079.45). By decreasing our confidence level, we make the confidence interval smaller. The following diagram illustrates all three of these situations:

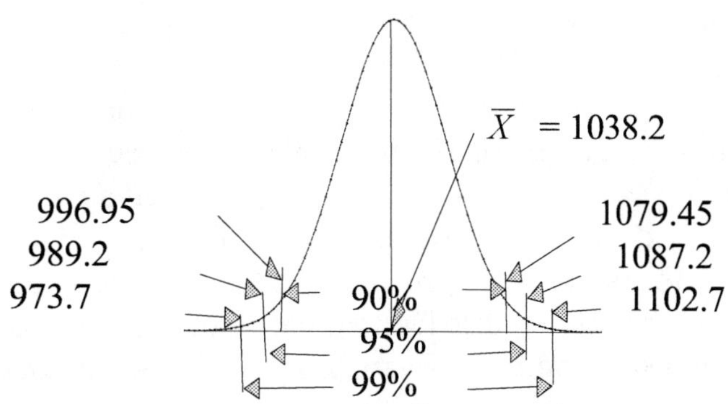

From the preceding example, we see that there is a direct relationship between the confidence level and the length of the confidence interval. If we increase our confidence level, we increase the length of the confidence interval. If we decrease our confidence level, we decrease the length of the confidence interval. Does this say anything about the precision or accuracy of our estimates? Since confidence intervals are built on sampling error, and sampling errors are measures of precision, confidence levels speak to the precision of estimates, not necessarily the accuracy. Recall from **DEFINITION 5.6.2**, precision is the absence of random error and accuracy is the absence of systematic error. If we have a larger confidence interval, we are more confident that our interval will contain the population mean and we are **more precise**. But by increasing the confidence interval, we are allowing for the possibility of more systematic error, i.e., we are decreasing the potential accuracy of the process. To increase the **accuracy** of our estimate, intuitively we need a smaller interval, i.e., we should **decrease** the confidence interval. But by decreasing the confidence interval, we have decreased our confidence that the mean is contained in that interval, and thus decreased the **precision** of our results.

In all of the preceding examples, we are dealing with large (according to **CONVENTION 5.6.1**) samples, i.e., samples with $n \geq 30$. What happens if we have a small sample ($n < 30$)? The Central Limit Theorem suggests that there might be a problem with such samples. However, there are times when it is impractical (or even impossible) to consider large samples. In certain industrial settings, using large samples presents serious problems. [What would it cost to use a large sample of Mercedes and test them for passenger safety in crashes? In this case, Mercedes Benz would like to be able to use as small a sample as possible.]

Very early in the twentieth century, William S Gosset, an English chemist and mathematician, working for Guinness Brewery, tried to find a way to save his employer money and himself time by using small samples to test the quality of the Guinness beer. In this attempt, Gosset developed a mathematical theory related to means and variances of small samples. Because this theory was developed "on company time" and "for the benefit of the company", Guinness claimed proprietary rights to the discovery and would not let Gosset publish his results. Guinness knew the "gold mine" on which they were sitting. What an advantage they would have if they could use small samples in their quality control processes while their competitors had to use large samples.

This concept of proprietary rights is quite current. Most employers require employees to sign a statement saying that anything developed which is related to their jobs belongs to the employer. This can involve a great deal of money in the case of some patents.

Gosset complied for several years, but finally decided that this result was of such importance that the "world needed to know it." Gosset published a paper reporting on his research under the pseudonym of Student. In this paper, he described a family of distributions, which he called the **Student t-distributions**. Each member of this family of distributions has the same general bell-shape curve as the normal distribution, but as one would intuitively guess, has greater variability than the normal distribution. This variability is dependent upon the sample size, i.e., the smaller the sample size, the greater the variability.

The following diagram illustrates the Student t-distribution for several values of n.

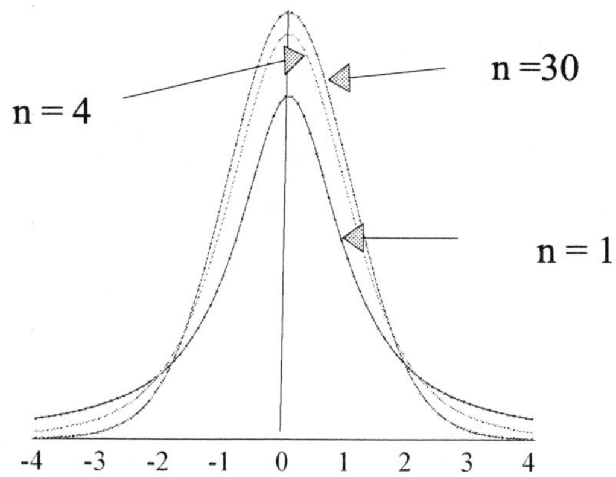

When do we use the t-distributions instead of the normal distribution? First, when we are dealing with small samples. Second, when we do not know the population standard deviation. The underlying population, however, must be essentially normal.

What happens if these assumptions are not true? If the underlying distribution is not normal but the population standard deviation is known, the Central Limit Theorem indicates that the sampling distribution still approaches the normal distribution for large n. If the underlying distribution is not symmetric, then there may be other methods (called non-parametric methods) which we can use.

The Student t-distributions have a solid mathematical basis. They were derived in response to the need to have something which worked for small samples. However, there is a Student t-distribution for every degree of freedom. Therefore, theoretically we can use the t-values for every degree of freedom. We use it for values of samples sizes less than 30, because for values of $n \geq 30$, the difference between the t-values and the z-values is insignificant. It is "not enough to worry about."

With the advent of the computerized statistical packages, the question of determining critical values is no longer a problem. We let the computer do the work. Many social scientists (and others) are now saying, "Let's use the t-distributions whenever we do not know the population standard deviation." Neither approach is wrong.

SECTION 9 - SUMMARY

In this chapter we introduced probability as a measure of the likelihood that a particular event will occur. We looked at various ways of expressing probabilities and considered two types of probabilities. We then considered the concept of a random variable and looked at examples of discrete and continuous random variables. We considered one particular continuous distribution, known as the normal distribution and looked at a number of applications of it.

We next looked at the Central Limit Theorem and the distribution of sample means. We calculated the standard error and used it as a measure of the precision in estimating the population mean by a sample mean.

We defined two types of estimates, point estimates and interval estimates. In moving from point estimates to interval estimates, we defined the concept of the sampling error and used it to construct confidence intervals. In working with confidence intervals for small samples, we introduced the concept of the Student t-distribution.

KEY WORDS AND CONCEPT

Probability
Classical or Theoretical Probability
Empirical or Experimental Probability
Random Variable
Discrete Random Variable
Continuous Random Variable
Probability Distribution
Normal Distribution
Standard Normal Distribution
Central Limit Theorem
Sampling Distribution
Standard Error
Accuracy
Precision
Point Estimate
Interval Estimate
Sampling Error
Confidence Interval
Confidence Level
Student t-Distributions
Degrees of Freedom

KEY FORMULAS

TERM	FORMULA
Standard Error of the Mean	$\sigma_{\bar{x}} = \frac{\sigma}{\sqrt{n}}$
Estimated Standard Error of the Mean	$s_{\bar{x}} = \frac{s}{\sqrt{n}}$
Confidence Interval for Large Samples	$(\bar{x} - z_{\alpha/2}\sigma_{\bar{x}},\ \bar{x} + z_{\alpha/2}\sigma_{\bar{x}})$

CHAPTER 6
HYPOTHESIS TESTING

SECTION 1 - INTRODUCTION

This chapter discusses the construction and testing of hypotheses. In Section 2, given a problem, we will construct two competing hypotheses, one of which must be true. Since the basis of the Central Limit Theorem is probability theory, we know that while we may judge a hypothesis as reasonable, there is some chance that it is not correct. We may be making an error. In Section 3, we will discuss the types of errors that are possible. In Sections 4 and 5, we discuss various types of tests to determine whether given hypotheses about the mean of a distribution are reasonable.

SECTION 2 - TYPES OF HYPOTHESES

How do we begin to solve a problem? One good way is to come up with possible solutions or outcomes. We make educated guesses about what these solutions could be. We call these guesses hypotheses, from a Greek word that loosely translated means "to suppose." In working with hypotheses, we will normally follow a procedure similar to the one outlined below.

Many of you have or used to have pocket calculators that use small nickel cadmium cells. Those cells have to be recharged after a certain number of hours of operation. Suppose the instructions suggest that the batteries can operate for 50 hours before they need to be recharged. We can ask the question, "Will these batteries really last 50 hours?" We could try to answer this question by running the batteries down several times and recording each time how long it takes. Suppose we ran the batteries down twenty times. From the previous chapters, we know that the probability that the mean life expectancy of the twenty trials equals exactly 50 hours is almost nil. However, we also know that if the population mean is 50 hours, then our sample mean will be close to 50. If our sample of 20 trials yields a sample mean of 58 and a sample standard deviation of 10, how likely is it that the population mean can be 50? We could illustrate the problem with the following sketch:

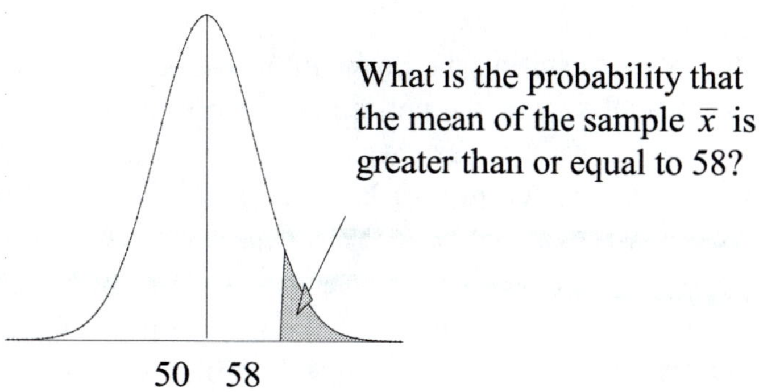

What we are really asking is whether the difference between the sample mean and the assumed population mean is due to random error or "chance," or is the assumption concerning the population mistaken. Will sampling ever provide enough evidence for us to conclude something absolutely about a population? In sampling, we will never "sample" all elements from a population. Thus we will never really "know" the entire population or the parameters from the population. What can we know? In reality, sampling and subsequent analysis can only provide enough evidence for us to say when a sufficient difference in our observations implies that there is a corresponding difference in the population.

How do we know when we have sufficient evidence? We will use a method known as hypothesis testing. It is almost like a game of "let's pretend." We start with an assumption concerning a population parameter. This assumption is known as the **NULL HYPOTHESIS**. We then look at our sample and decide how likely it would be for our sample to come from a population with the given parameter. Note that hypothesis testing is really not "proving" the null hypothesis. Rather, it is showing how likely it is that the null hypothesis is true or false. If we come to the conclusion that our null hypothesis appears to be false, then we in effect come to the conclusion that an **ALTERNATIVE HYPOTHESIS** appears to be true.

To make sense of this process of hypothesis testing, there are several properties of our competing hypotheses that make the process work. First, the two hypotheses cannot both be true simultaneously. If the null hypothesis is true, the alternative cannot be true. If the null hypothesis is false, then the alternative cannot be false. We say that the hypotheses are **MUTUALLY EXCLUSIVE**. Second, we assume that one of the hypotheses must be true. The two hypotheses cover all possibilities. We say that they exhaust all possibilities, or they are **EXHAUSTIVE**. If we combine these two properties, we see that one hypothesis must be true and the other false. To formalize this procedure, consider the following example:

EXAMPLE 6.2.1: The national average for the SAT Verbal score in a recent year was 427. This average was based on the scores from 1,088,223 test takers. If the freshman class of 556 students at Ivy Hall University had an average Verbal SAT of 499, was this freshman class significantly better than the national average?

Another way of stating the problem above is the following: "Was the freshman class at Ivy Hall representative of the entire population that had a mean of 427; or was it drawn from a subpopulation whose mean was greater than 427?" To begin looking at this problem (and the process), we need to translate the problem into a form with which we can work. In this case (as in most cases) we will translate the problem into mathematical or logical symbols. Let us assume that the entire population mean can be represented by the statement $\mu_0 = 427$. If we let μ represent the mean of the population from which the freshman class of Ivy Hall University was drawn, then the question that we are trying to answer is the following: "Is $\mu > \mu_0$?" Thus, one hypothesis to be considered would be

$$H: \mu > \mu_0$$

From our previous discussion, the proper alternative hypothesis must be mutually exclusive and exhaustive. The only hypothesis which meets those two criteria is [Why?]

$$H: \mu \leq \mu_0$$

In our previous discussion, we have noted that sampling will not "prove" something about the entire population. The best that it can do is to show us how likely or unlikely it is that a given hypothesis is true. With that in mind, we find that it makes more sense to state the null hypothesis in terms of equality, or at least include the equality possibility within the null hypothesis. Thus, for our Ivy Hall example, we have the null hypothesis

$$H_0: \mu \leq 427$$

and the alternative hypothesis

$$H_1: \mu > 427.$$

In **EXAMPLE 6.2.1**, we wanted to show that the sample did not come from the population under consideration. This problem represents a wide class of research problems in which we will be looking for evidence to conclude the sample has a characteristic that is new or different from the known population. In the example above, we formulated what we wanted to know as our alternative hypothesis (H_1: $\mu > 427$). As a general guideline, we will do this in all research-related problems. The hypothesis in which we are interested (research hypothesis) should be formulated as the alternative hypothesis. Thus, if we reject the truth of H_0 and therefore infer that H_1 is true, we will be concluding what we want to conclude.

Another type of problem involves testing the validity of assumptions. Suppose a firm that offers a program to help high school students prepare for the SAT's claims that their methodology guarantees an increase of at least 100 points on the combined verbal and math scores. How would we test such a claim? We could select a sample and do pre- and post-testing to see if the firm's claim is accurate. This is the same type of rationale used in criminal trials. The accused is assumed innocent until proven guilty. The firm is assumed to be telling the truth until proven otherwise. If we let μ be the mean improvement between the pre- and post-tests, the hypotheses would be stated as follows:

$$H_0: \mu \geq 100$$
$$H_1: \mu < 100$$

If the sample suggests that the null hypothesis cannot be rejected, then the firm's claim cannot be challenged. In any problem that involves the testing of the validity of a claim, the null hypothesis is usually formulated based on the claim or assumption. The rejection of the null hypothesis will provide the evidence that the claim is inaccurate or incorrect.

A third type of problem involves the choice between two courses of action. Suppose you use 2-inch bolts in a fabrication of camping trailers. Because of how the bolts are used, the bolts cannot be much shorter or much longer than 2 inches. The bolts are purchased in bulk. How will you know whether a shipment of 2-inch bolts is good? We can select a sample of bolts and formulate our hypotheses as follows:

$$H_0: \mu = 2$$
$$H_0: \mu \neq 2$$

If the "average" bolt is greater than or less than 2 inches, we should reject the null hypothesis and return the bolts to their manufacturer. If the sample suggests that H_0 cannot be rejected, we should accept the shipment of bolts.

If the two hypotheses must be exhaustive and mutually exclusive, then in examples involving the mean of a sample and a mean of a population, the examples above are the only three appropriate combinations of null and alternative hypotheses. If we let μ_0 be the value of the mean in which we are interested, then the three possibilities are the following:

I. $H_0: \mu = \mu_0$
$H_1: \mu \neq \mu_0$

II. $H_0: \mu \leq \mu_0$
$H_1: \mu > \mu_0$

III. $H_0: \mu \geq \mu_0$
$H_1: \mu < \mu_0$

Thus, hypothesis testing is an attempt to determine which of two competing hypotheses are most consistent with the sample data. If we find that the evidence provided by the sample cannot support the null hypothesis, we will **REJECT** that hypothesis and **ACCEPT** or **RETAIN** the alternative hypothesis. By rejecting the null hypothesis, we are not saying that the null hypothesis **IS FALSE**. We are saying that the evidence **SUGGESTS** that the alternative hypothesis is true. The results of hypotheses testing cannot be taken as absolute certainty.

If we look at the three sets of hypotheses above, we see that there is a fundamental difference between set I and sets II and III. Set I involves a straight equality statement. The alternative hypothesis involves two areas, one area above the given value and one area below it. In Sets II and III, the alternative hypothesis of each involves one area, either above or below the value under consideration. Set I involves two sides or tails of a distribution. It will be called a **TWO-TAILED TEST**. Sets II and III each involve one side or tail of a distribution. They will be called **ONE-TAILED TESTS**. We can go even further in our description. Since the alternative hypothesis in Set II is concerned with an area above the mean, we can call this an **UPPER-TAIL TEST**. Similarly, since the alternative hypothesis in Set III is concerned with an area below the mean, we can call this a **LOWER-TAIL TEST**. The "direction" of the inequality in the hypothesis H_1 in Sets II and III will determine whether we should be using an upper- or lower-tail test.

We will use two-tailed tests when the results could be on either side of our given value or when we do not know on which side our sample results will fall. If we construct our hypotheses before we select a sample, we will not know whether the sample mean will be greater than or less than the population mean. If we do have some indication as to the direction in which we are interested, then we can use an upper-tail or lower-tail test.

SECTION 3 - TYPES OF ERRORS AND LEVEL OF SIGNIFICANCE

Given a null hypothesis and its alternative hypothesis, how many true-false combinations exist? There are four possible scenarios. We know that the null hypothesis must be either true or false. Given the truth or falsity of the null hypothesis, we can either accept the null hypothesis or reject it. The first possible scenario is the acceptance of a null hypothesis that is true. In doing this we would be making a "correct decision." This is a "good" situation. The second scenario is the rejection of a null hypothesis that is true. By doing this we would be making a mistake or committing the error of accepting a false hypothesis. This is a "bad" situation. The third possibility is to reject the null hypothesis when it is true. To do this we are making a mistake. This also is a "bad" situation. The final possibility is to reject a false null hypothesis. This would be a correct decision and a "good" situation. We can diagram these possibilities as follows:

TABLE 6.3.1

		State of Nature	
		The null hypothesis is true	The null hypothesis is false
Decision	We accept the null hypothesis	Correct decision	Type II Error (Type β)
	We reject the null hypothesis	Type I Error (Type α)	Correct decision

In the diagram we have labeled the two errors Type I (or Type α) and Type II or (Type β). A **TYPE I** error is the mistake of rejecting the null hypothesis when it is true. A **TYPE II** error is the mistake of failing to reject a null hypothesis that is false. An obvious question is: "Which type of error is worse?"

Suppose we have a vaccine that we know works well, but a pharmaceutical firm has developed a new vaccine. Suppose we know that the cost of the two vaccines is the same, and the new vaccine has no additional side effects from those exhibited by the first vaccine. Thus, cost and side effects should not be factors in our decision of which vaccine to use. The decision of which vaccine to use should be based solely on the relative effectiveness of the two vaccines.

How do we formulate a null hypothesis? Suppose we say that the new vaccine is, at its best, only as effective as the old vaccine. Let μ_{new} be the mean effectiveness of the new vaccine and μ_{old} the mean effectiveness of the old vaccine. Given the assumption that the new vaccine is at its best, only as effective as the old vaccine, an appropriate null hypothesis can be written as

$$H_0\text{: } \mu_{new} \leq \mu_{old}$$

Since we have a working vaccine, we should not shift away from it unless the new vaccine is clearly better. A Type I error would be the rejection of a null hypothesis that is true. Now, if the new vaccine was at its best only as effective as the old vaccine, we would not want to "jump" to the new vaccine, i.e., accept the alternative hypothesis that the new vaccine was better than the old one. Committing a Type I error could be exposing people to great medical risk by not providing the best care possible.

But if the new vaccine is clearly better, we should shift away from the old vaccine. If the new vaccine is clearly better, and we do not use it, we are not giving people the best medical care possible and exposing them to risks that they should not have to take. We are committing a Type II error that also has serious consequences. Both types of errors have very serious consequences. It is very difficult to say which is more serious.

Suppose we are involved in a criminal trial and our null hypothesis is that the accused is innocent. We do not want to convict an innocent person, so we do not want to make a Type I error (rejecting a null hypothesis that is true). In this case, committing a Type I error has serious consequences. However, in our desire to protect the rights of individuals (by decreasing the probability of committing a Type I error), we will be increasing the likelihood of making a Type II error. In general, the probabilities of making Type I and Type II errors are inversely proportional. As we decrease the likelihood of one type of error, we are increasing the likelihood of making the other type of error.

How can we live with such a dilemma? To understand the situation, let us review the process again. Our objective is to make inferences about a population knowing only information from a sample. Can we ever prove, or even disprove, a null hypothesis? No! What we can do is come to a reasonable conclusion by using probability theory. What is reasonable? To define what is reasonable, we need to make a judgment. How willing are we to accept something occurring by chance? Are we willing to reject a null hypothesis based on evidence that can happen by chance 5% of the time? If not, are we willing to reject that hypothesis based on evidence that could happen by chance 1% of the time? By selecting such a value, we are in reality setting a level at which we are willing to make a Type I error. We call this value the **α-LEVEL**, or the **LEVEL OF SIGNIFICANCE**. The value α is the probability of making a Type I error, given that H_0 is true.

Although we could choose any α, there are two typical α-levels, $\alpha = 0.05$ or $\alpha = 0.01$, which are used in most statistical work. On occasion other values of α may be used depending upon the seriousness of making a Type I error. Once we have decided the level of significance, the probability of making a Type II error will be dependent upon that level of significance as well as the population or sample size. This probability, denoted by β, will be inversely proportional to the level of significance and to the sample size. For a given level of significance, the larger the sample size, the smaller β will be. For a fixed sample size, the larger the level of significance, the smaller β will be. The actual determination or calculation of β is beyond the scope of this course.

In practice, instead of looking at the value β, the value $(1 - \beta)$ is used. This value represents the probability that we will reject H_0 given that H_0 is false, or in other words, the probability of accepting H_1 given that H_1 is true. The value $(1 - \beta)$ is called the **POWER OF A TEST**. The greater the value of $(1 - \beta)$, the more powerful the test, i.e., the more likely we will reject the null hypothesis when it is false. The following example illustrates Type I and Type II errors:

EXAMPLE 6.3.1 A research company dealing with highly classified matters can grant access to a high security area of a research laboratory only to authorized personnel. The company installs a "palm reader" to guard the area. The palm reader will open the one door to the area only if the reader can identify the individual's palm print. The palm print of every individual who can enter the area is programmed into a computer. A person who wants to enter the area will place his or her palm on a reader plate by the door. If the computer matches that print with a print in its memory, the door opens. If the computer does not match the print, the door will not open and the company's security force is immediately notified that an unauthorized access is being attempted and dispatched to the area where access is being attempted. In effect, this system is doing hypothesis testing:

H_0: The individual attempting access is authorized.
H_1: The individual attempting access is not authorized.

Because of extensive prior testing, the palm reader's manufacturer claims that the Type I error rate is less than 0.1%, while the Type II error rate is 0.00001%. A Type I error is the rejection the null hypothesis when it is true. In this situation, what is a Type I error? A Type I error would involve denial of access to an authorized individual. What does a Type I error rate of less than 0.1% imply? It implies that an authorized individual will be denied in fewer than 1 out of 1000 attempts at access. Such an individual will most likely only be inconvenienced for a short time. When the security force arrives, the individual can identify himself or herself, "reapply" his or her palm, and almost assuredly gain access to the lab.

A Type II error is the acceptance of a false null hypothesis. In this case, it would be the granting of access to an unauthorized individual. What does a Type II error rate of 0.00001 imply? It implies that unauthorized access will be granted to individuals only 1 time in 10 million. To understand the magnitude of this, suppose the palm reader takes 30 seconds to analyze and verify a palm. The reader can do 120 palms per hour, or 2880 per day maximum. In one year, it could read a maximum of 1,051,200 palms. At this rate, it would make one Type II error every nine years or so. But, this assumes that it is in constant and continual use. It is very unlikely that this would be the case. Generally, such a mechanism would only be used on areas with a few people coming and going. If instead of 120 attempts per hour there were only an average of 10 attempts at access per hour, the mechanism would make one Type II error about every 95 years.

Thus, it is very unlikely that the palm reader will make a mistake and allow an unauthorized individual access. However, such a security system is not without flaws. The system does not allow for the "unauthorized" use of an authorized individual's palm. Such a *terrorist* approach would be difficult to stop. If the company felt that this was a possibility, other counter measures could be taken. There exist even more sophisticated control mechanisms, such as "eye readers," which check the unique pattern of blood vessels across the retina. In several recent movies and television shows, it has been suggested that such readers are in use by the United States military to check the individuals who can launch nuclear warheads. Several ways "to fool" the eye reader were also suggested.

Thus, so far, we have looked at two parts of the hypothesis testing process. First, given a problem, we formulated an appropriate null hypothesis H_0 and an appropriate alternative hypothesis H_1. Also, based upon the perceived seriousness of a Type I error for the given problem, we selected an appropriate level of significance α.

What do we do now? In our formulation of appropriate hypotheses, we noted that we will be looking at population parameters such as the mean of the population, and comparing those parameters with sample statistics, such as the mean of a sample. How will we make those comparisons? We will need some standard against which we can compare our sample statistics. What standard or yardstick can we use? The previous chapter's discussion of the standard normal distribution and the Central Limit Theorem provides us with some direction. In our discussion of confidence intervals we noted that there is an implicit test statistic associated with each given problem or situation. On the basis of our hypothesis, we select a procedure to test the hypothesis.

The process works the same for any scenario involving hypothesis testing:

1. **Hypotheses**
   State H_0: null hypothesis and H_1: research hypothesis
2. **Significance Level**
   Choose α (the most common level is .05)
3. **Test Statistic** (also called calculated value)
   Choose the appropriate statistical test and perform it (usually on the computer)
4. **Conclusion**
   Compare the p value to α
   If $p \geq \alpha$, retain H_0. If $p \leq \alpha$, reject H_0
   Express the decision in the context of the problem

The p value is the probability of obtaining the test statistic by chance and chance alone, given the H_0 is true. Stated another way, p is the smallest value at which H_0 can be rejected for the given data. So if the significance level is set at α=.05 and the p value is determined to be p=.31, it means that the probability of the sample data occurring by chance is 31%. Since we said we would reject H_0 only if our sample could occur less than 5% of the time, we must retain H_0. Our sample is not "rare enough" to reject H_0 and therefore be statistically significant.

For problems dealing with large samples from a distribution that is essentially normal and for which the standard deviation is known, we can use a z-score as a comparison or test statistic. This will be known as a **z-test**. The next section deals with z-tests and when they are appropriate and how to use them. The next section will also provides us with the pattern we will use for all hypotheses testing problems.

SECTION 4 - LARGE SAMPLE TEST OF MEAN

In the previous sections of this chapter we set the stage for the process of hypothesis testing. In this section, we will look at the particular test statistic known as the z-test. We will indicate when this test is appropriate and how to use it. We will also develop the general hypothesis testing procedure using the z-test as our model. To illustrate the test, reconsider the Ivy Hall freshman class that was introduced in EXAMPLE 6.2.1.

EXAMPLE 6.4.1: The national average for the SAT verbal score was 427, based on the scores from 1,088,223 test takers. If the freshman class of 556 students at Ivy Hall University had an average verbal SAT of 499, was this freshman class significantly better than the national average?

What do we do next in the process? The question we want answered is the following: "Is the freshman class at Ivy Hall better than the 'average' freshman nationally?" From our previous discussion, we propose the following null and alternative hypotheses:

H_0: $\mu \leq 427$
H_1: $\mu > 427$

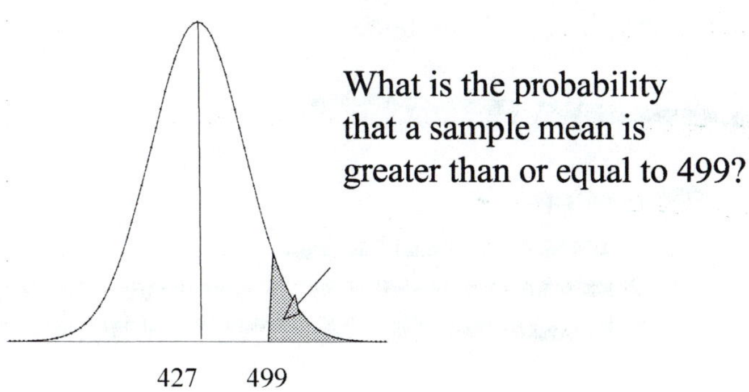

The "sample" of Ivy Hall freshmen has a sample mean of 499, and is pictured above. Thus we are "testing" the sample mean. The Central Limit Theorem tells us that the distribution of sample means can be approximated by a normal distribution. The critical question is whether 499 is sufficiently above 427 to make a difference? Also, since we are dealing with a large sample of 556 students, the Central Limit Theorem allows us to compare our results with the normal distribution using z-scores.

Given an α-level of 0.05, we need to know if 499 occurs less than 5% of the time by chance. From the direction of the inequality in the alternative hypothesis H_1: $\mu > 427$ and the picture above, we also see that we are only interested in the area above the mean. Since we are interested in only one side of the mean, this test is called a **ONE-SIDED TEST**. Since we are interested in the region above the mean, it is an **UPPER-TAIL TEST.** Our significance level $\alpha = .05$ sets up the upper tail area of 5% to be the rejection region. We will only reject the H_0 if our sample result (test statistic) falls in the rejection region (the least likely and largest 5% of all sample results). This is the same as saying the p value must be less than .05. If our sample result (test statistic) falls in the acceptance region, it is the same as saying $p>.05$. This can be illustrated graphically below:

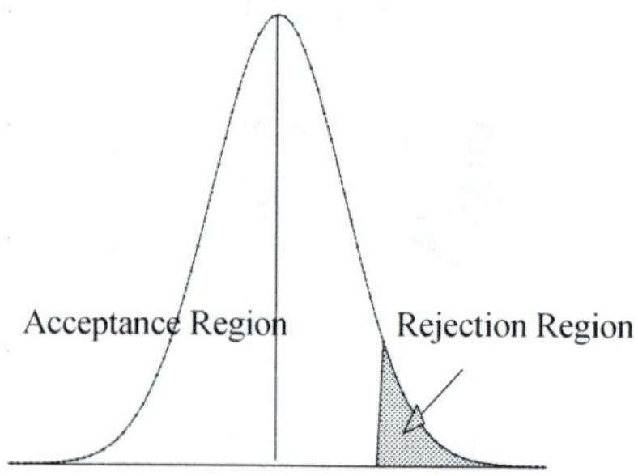

We conduct the test on the computer, either selecting one-sample z-test or constructing a confidence interval around our sample mean and determining whether the interval contains the hypothesized mean. In this case, there is sufficient evidence to conclude that the population from which the freshman class for Ivy Hall University was better than the entire population of verbal SAT test takers.

Let's review the hypothesis testing process:

1. State H_0 and H_1.
2. Set the significance level α.
3. Choose the appropriate statistical procedure and perform it.
4. Compare the p value to α and come to a decision.

It is extremely important to note several points about this process. The first is that the first three steps must be completed before we can go on to the next step. We must select our hypotheses and level of significance before we do any calculations. It is "unethical" to calculate the test statistic and view the p value before selecting the level of significance. [Why?]

The second point is that when we translate our problem into hypotheses, there are some general guidelines that we can [and should] use. Most of the time a problem will naturally ask a question or suggest a claim. "Is this freshman class better than previous ones?" is a question. This question can be naturally translated into a claim: "This freshman class is better than previous freshman classes." If we were to represent this claim by a mathematical statement, it would contain a "greater than" sign in it. It would not contain an equality statement. You will recall that we said that our null hypothesis should either be an equality statement or contain an equality possibility (=, ≤ or ≥). In this case, our original claim will be our alternative hypothesis. Each problem will ask a question or make a claim. As we translate these questions or claims into hypotheses, we need to remember to include the equality statement in the null hypothesis and determine whether our original question or claim corresponds to the null or alternative hypothesis.

The next point concerns the meaning of acceptance or rejection of the null hypothesis. When we reject the null hypothesis, we usually say that we accept the alternative hypothesis. We have already noted that acceptance of the alternative hypothesis does not prove that the alternative hypothesis is true. It suggests [strongly suggests] that the alternative is true, but does not absolutely prove it.

What happens when we do not reject the null hypothesis? We have said that in this case we accept or retain the null hypothesis. Does this prove that the null hypothesis is true? NO! Does this suggest that the null hypothesis is true? Not necessarily? Why the difference accepting the null hypothesis and accepting the alternative? Because of how we set up the hypotheses, our **ACCEPTANCE REGION** is the larger region, while the **REJECTION REGION** is usually much smaller. If the value of our test statistic falls in the rejection region, it is highly likely that our sample represents a population different from the population represented by the null hypothesis.

For the above reasons, many authors do not like [and will not use] the phrase **accept the null hypothesis**. These authors say that the evidence **fails to reject the null hypothesis**. What is suggested by this terminology is more like "There just is not enough evidence at this time to reject the null hypothesis." In this text we will use the phrase **accept or retain the null hypothesis**. However, you need to be aware of the implications of this.

The final point we need to make is that we have not completed a problem until we have translated the mathematical or statistical results back to that original problem. If we want to know whether a particular freshmen class is better than all college-bound students, it is not sufficient to "Reject the null hypothesis." As you get into doing problems, you will see that there is not necessarily only one correct way to formulate null and alternative hypotheses. Thus, one person's acceptance region could have some close correspondence to another's rejection region. We need to apply our answer back to the original problem and state the "solution" in terms related to the problem.

Let's look at another example of the hypothesis testing process:

EXAMPLE 6.4.2: A local electric company, Metropolitan Edison, known as Met Ed, has asked the Public Utilities Commission for a rate increase. Met Ed suggests that part of its rationale for the rate increase is the fact that its residential customers are using more electricity this year than they did previously. [It is more complicated than it first appears, but if you sell more, you can raise your prices to make more profits. If you are not selling enough, you lower your prices to sell more.] To support its claim, Met Ed suggests that historically, the mean consumption of electricity for the month of December has been 1753 kilowatt-hours per residential customer with a standard deviation of 185.5, while for a random sample of 60 residential customers the mean consumption for the current month of December is 1802.8 kilowatt-hours. Does this information imply that the mean consumption has indeed increased?

Using the method outlined we have the following:

1. Formulate appropriate null and alternative hypotheses:

   For this problem the utility company has recorded a historical pattern of a mean of 1753 kilowatt hours of electricity usage per residential customers for the month December. For a given sample for the most recent December, the company found that the sample mean was greater than the historical mean. The company would like to conclude that usage has increased. If we formulate our alternative hypothesis as the hypothesis in which we are interested, we would say that the new mean "is greater than 1753". The corresponding null hypothesis then would be that the mean "is less than or equal to 1753." Thus, we have the following hypotheses:

   H_0: $\mu \leq 1753$
   H_1: $\mu > 1753$

2. Select an appropriate level of significance α:

   Since the utility company is attempting to get a rate increase, we would like to be very certain that usage has indeed increased before we allow the rate increase. [We are trying to protect the public.] Therefore, let us assume $\alpha = 0.01$.

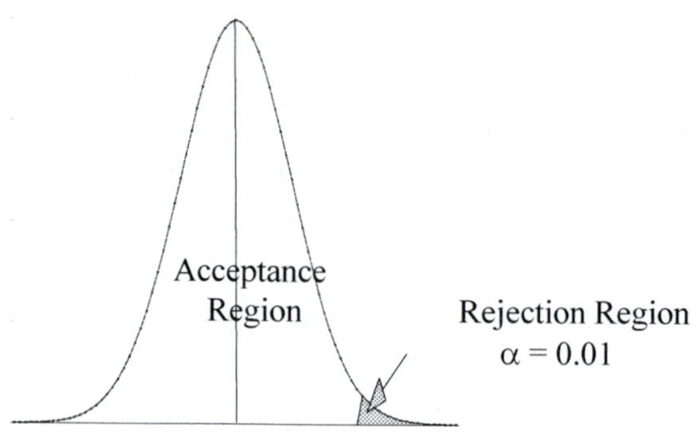

3. Choose the appropriate statistical test and perform it.

   Using a one-sample z-test on the computer, our sample mean of 1802.8 gives a p value of 0.0188.

4. Conclusion

   Since our $p > \alpha$ (0.0188 > .01), we must retain the null hypothesis. This means that there is not enough evidence to support the utility company's claim that residential consumption has increased. We should not grant the utility company's request for a rate increase.

The above example illustrates an important reason for determining the level of significance before you actually calculate the value of the test statistic. What would happen if we selected the level of significance $\alpha = 0.05$? Since our p value turned out to be .0188 < .05, we would reject the null hypothesis and accept the alternative. This says that usage has increased. Therefore, the utility company would have justification for their rate increase. In one case we rejected the null hypothesis and in the other we accepted it. Does that mean that statistics lie, or are at best meaningless? **NO!** However, it does illustrate why the level of significance should be selected before any calculations are done. To calculate the test statistic and then select a level of significance that would give us a "desired" result would be an unethical use of hypothesis testing. We should not participate in such a practice, and when we see such a procedure in use, we should question it.

Can the hypothesis testing procedure be simplified any? **NO!** Every step in this process is essential. However, we can simplify the reporting or recording of the process (as long as we remember we are not short cutting the process). To illustrate, consider the following example of a z-test.

EXAMPLE 6.4.3: On a standardized IQ test, the scores are normally distributed. The mean for adults (individuals aged 18 or older) is 100 with a standard deviation of 15. A psychologist wants to know whether, using this IQ test as a measure, prisoners within the federal penitentiary system are "more intelligent" or "less intelligent" than the "average" adult.

From the given information, she knows that the mean intelligence μ of the entire adult population is 100, while the standard deviation σ for this distribution is 15. However, she does not know whether federal prisoners score better or worse on this IQ test.

Since the psychologist is asking whether federal prisoners are "different from the general population, she should use a two-sided hypothesis test. The mean of the general population is 100, so she forms the following hypotheses:

1. Hypotheses

H_0: $\mu = 100$ (prisoners have the same intelligence as an average adult)
H_1: $\mu \neq 100$ (prisoners have a different average intelligence than an average adult)

2. Significance level

Since this is not a matter of "life and death," the psychologist selects α=.05 as her level of significance.

3. Test Statistic

Using a one-sample test on a sample of 50 individuals with a resulting sample mean of 106.2, the computer gives us p=.0018.

4. Conclusion

Since $p<.05$, the psychologist must reject the null hypothesis and accept the alternative hypothesis. She concludes that federal prisoners have a different average intelligence than the general population.

SECTION 5 - SMALL SAMPLE TEST OF MEAN

We have shown how to do hypothesis testing for situations in which we have a large sample and are dealing with a population with a known standard deviation. In these cases, we may use the z-test. There are three questions that could be asked at this point: (1) What is a large sample? (2) What happens if we do not have a large sample? and (3) What happens if we do not know the population standard deviation σ? If the sample size is less than 30 or if the population standard deviation is not known, the z-test is replaced by another test called the **t-test.** This test uses the **Student t-distribution** defined in the previous chapter. The following examples introduce the test and illustrate some of its uses.

EXAMPLE 6.5.1: A nationally advertised diet program claims that "by drinking its delicious milk shakes for breakfast and lunch, and by eating a sensible dinner, a person could lose 20 pounds in a month." To test the truthfulness of this ad, a consumer group asks 26 "overweight" volunteers to "faithfully" follow the program for a month. At the end of a month, the dieters lost the following weight:

20.1	18.7	15.9	15.6	23.7	19.7	23.5	24.2	17.8
21.4	19.8	22.4	7.6	24.3	23.5	22.1	26.7	18.8
23.4	20	14	20.8	9.5	25.2	16.9		

Is the ad misleading? Suppose we go through the hypothesis testing process to answer the question:

1. Hypotheses

What are we after? The main question asked is "Is the ad misleading?" How can we interpret that? What is the primary claim made by the company? "One can lose 20 pounds in a month by using the plan." What would be a reasonable interpretation of that claim? If one person out of all people that ever used the diet plan lost 20 pounds, then one could say that "One can lose 20 pounds." However, most reasonable people would not agree this is acceptable advertising. What would be an acceptable interpretation of the claim?

If everyone who tries the diet plan loses 20 pounds, then the company is shortchanging itself in the ad. The ad could be much stronger. Is there another interpretation that fits the situation? What about an average weight loss of 20 pounds? If the average weight loss per person is 20 pounds, then an ad that claims that a person can lose 20 pounds would not be construed to be misleading. Thus, we will assume the ad means that the average weight loss per person is 20 pounds. Another way of saying this is the following: "The mean of the distribution of weight losses is 20 pounds." Symbolically, $\mu=20$.

What information does the sample of 26 people give us? How do we fold that information into the mix? We can easily find the mean weight loss of these 26 people and compare that number with the "reported" mean of 20 pounds. Our problem then becomes, "Is the mean weight loss of the sample sufficiently different from 20 to bring into question the claim of the ad?"

Another way of stating this problem is the following: Let μ_o be the mean of the distribution from which our sample is taken. Is μ_o sufficiently different from μ to lead us to believe that the distribution from which the sample is taken is different from the population used to arrive at an average weight loss of 20 pounds for the diet program?

Since the primary question concerns whether two means are sufficiently different, we should write our null hypothesis in terms of an equality, i.e.,

H_0: $\mu = 20$ (average weight loss is 20 pounds)

The alternative hypothesis then becomes

H_1: $\mu \neq 20$ (average weight loss is not 20 pounds)

We should note that this alternative hypothesis points to regions in two different directions or sides of the hypothesized mean. This is a **TWO-SIDED TEST.**

2. Significance level

Although dieting can be a matter of life and death in certain cases, the question of whether an ad about a dieting program is misleading is not of cosmic importance [cosmetic importance, yes!]. However, if the company that publishes the ad could be prosecuted, the company would want the evidence to be "overwhelming" to bring in a decision against them. This really means that we could select either a significance level of $\alpha=.05$ or $\alpha=.01$ with justification. Suppose for our problem we select $\alpha=.01$.

3. Test statistic

Select the one-sample t-test on the computer using n=26, sample mean of 19.8, and sample standard deviation of 4.5. The computer then gives us p=.035.

4. Conclusion

Since $p>.01$, we retain our null hypothesis and thus we must accept this claim. By retaining the null hypothesis in this problem, what we are saying is that we do not have enough evidence to dispute the company's claim; therefore, at this time, we must accept it.

In the example above, we have both a small sample and an unknown population standard deviation σ. Although the t-distribution was originally developed to explain the sampling distributions for small samples, it may be used whenever the population standard deviation is unknown. The following example illustrates this.

EXAMPLE 6.5.2: Suppose you are a curriculum coordinator in a rural high school. You are interested in the critical thinking skills of the students in your school. You come across a test that purports to measure critical thinking skills. The test gives norms for groups of individuals such as high school drop outs, high school seniors, liberal arts college juniors, college graduates and medical school graduates, as well as for all American adults. You like the test so much that you administer it to all 121 students in your senior class.

When you score the test, you find that your seniors have a mean score of 46.2. The national high school senior subgroup was reported to have a mean score of 46.8. However, one problem that you notice in the testing material is that although the publishers have suggested a standard deviation for the entire population, they did not publish any standard deviations for subgroups. You want to know: "Are my students really less prepared in critical thinking skills than the national group of high school seniors?"

1. Hypotheses

Since the primary concern is whether your seniors are less prepared in critical thinking than the national average, we can use a one-tailed test of hypotheses. Since you have some preliminary information that suggests that your seniors are not as well prepared, you should use the "less than" statement in your alternative hypothesis:

H_0: $\mu \geq 46.8$
H_1: $\mu < 46.8$

2. Significance level

You know critical thinking skills are very important to individuals, and helping one individual to acquire better skills could change the course of human history, or at least local history. However, you also know that measuring critical thinking skills is at best an imprecise activity. Thus you make a judgment that you can live with a .05 level of significance.

3. Test statistic

Select the one-sample t-test on the computer using data collected from the sample of n=121 and a sample mean and standard deviation of 46.2 and 1.2 respectively. On the computer we get $p<.005$.

4. Conclusion

Since $p<.05$, we reject the null hypothesis and conclude that the students are not as well prepared in critical thinking skills as students generally.

SECTION 6 - SUMMARY

In this chapter we introduced the concept of hypothesis testing and a process for testing hypotheses. We discussed the types of errors which can be made in testing hypotheses, levels of significance, the selection of an appropriate level for a given test, how to choose an appropriate test statistic and interpret its p value given by the computer, and how to make a conclusion based on the preceding steps of the process.

KEY TERMS AND CONCEPTS

Hypothesis Testing
Null Hypothesis
Alternative Hypothesis
Type I Error
Type II Error
Level of Significance
Test Statistic
One-tailed Test
Upper-tailed Test
Lower-tailed Test
Two-tailed Test
Rejection Region
Acceptance Region
p value

CHAPTER 7
HYPOTHESIS TESTING FOR TWO POPULATIONS OR SAMPLES

SECTION 1 - INTRODUCTION

In the previous chapter, we introduced the concept of hypothesis testing with examples which used only one population or sample. There are many situations, however, where you compare two samples and ask the question, "Do these two samples come from different populations or could they come from the same population?" A manufacturing company has two plants producing the same product. Are there any differences between the products from the two plants? A canning operation has two machines. Are there any differences in the volume of the cans filled by the two machines? A diet plan guarantees a certain weight loss. If you weigh people before they begin the plan, and then after they have completed the program, is the company's claim valid? In this chapter, we will consider tests dealing with two populations or samples (both dependent and independent designs) along with the homogeneity of variance test.

SECTION 2 - TWO-SAMPLE DESIGNS

What are the similarities and differences among the following examples?

EXAMPLE 7.2.1: The owner/operator of an SAT score enhancement program advertises that her program will increase SAT scores by 100 points. How could we test the validity of her claim? One way to do so would be to select a group of students at random and administer an SAT exam. Then the operator of the program could "teach" these students how to get better scores. After the students complete her program, we could administer another SAT exam and compare the scores before the program and the scores after the program.

EXAMPLE 7.2.2: A doctor has developed a "holistic medicine" procedure involving diet, exercise, rest, and meditation for the treatment of high blood pressure. He claims that his procedure will lower blood pressure more drastically within two weeks than will the usual method of treatment using drugs. How can we test his claim? We can not use two treatments on the same individual, since we could never be sure if or how the two treatments interacted. We would have to set up two distinct samples, and give one sample the drug treatment and one sample the holistic medicine treatment. We could then compare the difference in mean blood pressure change.

EXAMPLE 7.2.3: An athletic trainer wants to determine if there are differences in the proportion of body fat between female athletes and male athletes. To help answer her question, she randomly selects a sample of female athletes and a sample of male athletes. She measures the body fat of each athlete, and calculates the mean body fat for the female and male athletes. When can we say that a difference between the means of the two groups is significant?

The most obvious similarity among the three examples is that they each involve two samples and corresponding sample means. What are some of the differences? In the first example, the same individuals are tested twice. In the second and third examples, we are dealing with samples of distinct individuals. In the third example, males are different from females. In the second example, we could not treat the same individual with both methods.

Depending upon how we construct the samples in the second example, there could be a real difference between that example and the third. In the second example, as we construct our samples we should keep several things in mind. We are testing whether two treatments have the same or different results. If we start out with "patients" that are already different, how will we know whether resultant differences were due to the treatment or to the original differences? Therefore, we need to construct our samples in such a way that they are "matched," at least with respect to variables which could influence the results. The patients in the two samples in this case should have similar blood pressures, and probably similar ages and weights.

In the first example, there is a definite relationship between the individuals in one sample and the other. They are the same individuals, tested before treatment and after treatment. This type of an experiment is called a **BEFORE-AND-AFTER TEST**. In the second example, when we match individuals in one sample with individuals in the second sample, there is also a relationship between such individuals. This type of test is called a **MATCHED PAIR TEST**. For both of these examples there is a relationship between the two samples, so we say they are **DEPENDENT**.

In the third example, there is no apparent relationship between the two samples. Even if we attempt to match individuals with similar heights, weights, ages and body build, there are anatomical and physiological differences between men and women which would preclude a relationship (mathematically speaking). When there is no relationship between the two samples in a given problem, we say that the two samples are **INDEPENDENT**.

It is unfortunate that we do not have different terms to use here. How many different ways have we used the terms **independent** and **dependent**? We will have to depend (oops) on the context to determine in which manner we are using these terms, and usually we can. We can summarize these definitions as follows:

DEFINITION 7.2.1

A **BEFORE-AND-AFTER DEPENDENT DESIGN** involves two measurements, taken on the same set of individuals. The first measurement is usually taken before a treatment is given and the second measurement after treatment is imposed.

DEFINITION 7.2.2
A **MATCHED PAIR DEPENDENT DESIGN** involves two samples in which the individual sample points are matched "pairwise," one element of each pair from both of the two samples, according to a set of criteria.

DEFINITION 7.2.3
Two samples are **INDEPENDENT** if the sample selected from one population has no effect on or relationship with the sample selected from the other population. If two samples are not independent, they are said to be **DEPENDENT.**

SECTION 3 - DIFFERENCE OF MEANS (DEPENDENT SAMPLES)

In this section we will consider a particular type of problem involving tests of hypotheses about the means of two dependent samples, restricting our attention to those problems where the underlying populations are normally, or are very close to being normally, distributed. Certain problem where the populations are not normally distributed can be solved but must use methods known as nonparametric statistics. Thus, we will consider problems which meet the following conditions:

1. We are testing a hypothesis about the means of two populations, from which we have drawn two dependent samples (of the same size).

2. We can assume that the two populations are normally distributed.

EXAMPLE 7.3.1 Suppose Green Taxi Co., a large chain of taxi cabs in the east coast, wants to test two brands of tires to determine whether there is any difference in wear. They test twenty tires, ten of brand F and ten of brand G, by assigning one of each to a car. After 30,000 miles the tires are checked for wear and the following data is obtained:

Car	1	2	3	4	5	6	7	8	9	10
Brand F wear (ml)	3.2	4.1	3.5	2.9	3.4	3.7	3.2	3.8	2.8	3.9
Brand G wear (ml)	3.1	3.9	3.7	3.3	3.6	3.8	3.0	3.6	3.1	4.1
Difference (F - G)	0.1	0.2	-0.2	-0.4	-0.2	-0.1	0.2	0.2	-0.3	-0.2

This is a matched-pair two-sample dependent design since one Brand F tire is paired with one Brand G tire *on the same car*. This allows better control of extraneous variables like weight, engine size, age, maintenance, road conditions, etc. In dealing with these samples, we could calculate the means and standard deviations of the two samples.

However, we can reduce our "work" somewhat and get a better measure of the difference in wear between the two tire brands by taking advantage of the relationship that exits between the two samples and finding the difference in wear between the two brands (by subtracting). If there turns out to be a difference between the two brands, there is a good chance the difference results from the brand of tire and not from one of the other extraneous variables. The hypothesis test follows:

1. Hypotheses

   H_0: $\mu_d = 0$ (the mean difference in wear between the two brands of tires is zero)
   H_1 : $\mu_d \neq 0$ (the mean difference in wear between the two brands of tires is not zero)

   These hypotheses form a two-tailed test.

2. Significance Level
   We select α=.05.

3. Test Statistic
   We select a two-sample dependent means test with a hypothesized difference of zero on the computer and get p=.3536

4. Conclusion

   Since p >.05, we cannot reject the null hypothesis. Therefore, there is not enough evidence to say that there is a difference in tire wear between Brand F and Brand G.

   Journal form: t(9)=-978, p>.05.

The **journal form** is the typical format for presenting statistical results in research. Because you will likely encounter it in scholarly journals, textbooks, and other research settings, we will use the journal form from now on as an additional method to report the conclusion. The journal form has a prescribed format:

> letter of the test (degrees of freedom)=test statistic, $p > \alpha$ if the calculated p value is greater than the significance level so that H_0 is retained or $p < \alpha$ if the calculated p value is greater than the significance level so that H_0 is rejected.

EXAMPLE 7.3.2: Suppose we have a claim that an SAT preparation course guarantees a gain of 100 points on the SAT for students who successfully complete the course. To test this claim we select 8 students at random. We administer an SAT pre-test to these students, put them through the prep course, and then administer an SAT post-test. Since it is a before-and-after experimental design, the two samples are of the same size [Why?] and are dependent. Since we are dealing with SAT scores, we may assume that the population(s) are normally distributed. The pre- and post-test results are given in the following table:

Student	A	B	C	D	E	F	G	H
Pre-Test	850	740	980	1020	970	1100	1230	1050
Post-Test	930	880	1060	1080	1100	1240	1350	1090
Difference (post - pre)	80	140	80	60	130	140	120	40

If we want to know whether the prep course can produce a consistent improvement of 100 points on the SAT's, we can look at the mean of the difference of the pre- and post-test and see if that difference is greater than (or equal) to 100.

1. Hypotheses

   The claim the prep course operator is making is that the pre/post difference is greater than (or equal) to 100 points. This claim would translate into the following hypotheses:

   H_0: $\mu_d \geq 100$
   H_1: $\mu_d < 100$

   These hypotheses form a one-sided, lower-tailed test.

2. Significance Level
   We select α=.05.

3. Test Statistic
   We select a two-sample dependent means test with a hypothesized difference of 100 on the computer and get p=.465

4. Conclusion

Since p >.05, we cannot reject the null hypothesis. Therefore, there is not enough evidence to refute the operator's claim that the score increase will be at least 100 points.

Journal form: $t(7) = -0.09$, $p > .05$.

SECTION 4 - HOMOGENEITY OF VARIANCE AND THE F-TEST

Before we can proceed to the two-sample independent design, we must consider another issue. For two samples to come from the same population, both the means and variances of the two samples must be "close enough" to each other and to the mean and variance of the population. So far in our hypothesis testing procedures, we have concentrated on means. As we look at the two-sample independent case, we will first look at the variances because the test for comparing two means is contingent upon the results of a comparison of the variances.

The title of this section uses a word which we have not yet introduced. The word **HOMOGENEITY** comes from the word **HOMOGENEOUS** which means "alike or similar." If you will recall in our discussion of the inferential statistical testing process, we do not prove or disprove statements; we measure the likelihood that they are true or false. In the test of **HOMOGENEITY OF VARIANCE** which we will use, we determine how likely it is that the two variances are the same. To introduce our test, consider the following example:

EXAMPLE 7.4.1: Have you ever been in a supermarket or a bank and faced the choice of selecting a line? Do you "always" pick the slowest line? Do you select the shortest line only to find that the person in front of you has five different transactions? A bank manager is very concerned with what seem to be many complaints she is receiving concerning waiting times in her teller lines. During a vacation in which she had to check into airports several times, she noticed that the airlines all used a single waiting line and customers were directed to the first open agent. The bank manager thinks this might work in her bank so she decides to experiment with the single-waiting-line concept. She takes a sample of waiting times on a Friday during the noon hour, the busiest time of the week at her bank. She asks each of her tellers to come in an hour early on Monday morning for an important meeting. Over the weekend, she installs "cattle shutes." At the tellers' meeting, she explains the concept and informs her tellers that they are going to implement the single waiting line immediately that Monday morning. That Friday [to give customers and tellers several days to adjust to the new arrangement], she takes another sample of waiting times during the noon hour. The results are listed below. Does the single waiting line make a difference in waiting times?

Waiting Times (in minutes)			
Individual Teller Lines		Single Waiting Line	
2.0	5.2	3.6	5.2
2.6	5.7	4.2	5.2
3.5	6.3	4.6	5.3
4.0	7.8	4.7	5.4
4.3	8.6	4.9	5.5
		5.0	6.4
$n_1 = 10$		$n_2 = 12$	
$\bar{x}_1 = 5.0$		$\bar{x}_2 = 5.0$	
$s_1^2 = 4.61$		$s_2^2 = 0.49$	

The bank manager sees that the single waiting time has made no apparent difference in the mean waiting time (both means are equal to 5.0 minutes). However, there is a large apparent difference in the variations of waiting times, as measured by the variances. It would seem that "on the average" the bank customers are waiting the same amount of time, but there are fewer extreme values. With fewer extremes (on the long) side, the bank manager will most likely get fewer complaints about the length of waiting times.

The preceding example indicates why we would be interested in testing the homogeneity of variances. We need to work on the how. We first consider the assumptions under which we can test variances. The first assumption is that we have two samples. There are different methodologies for testing one variance against a known value, and for testing the variances from more than two samples. In this chapter we are only considering tests of two variances. The second assumption is that the two samples are independent. In our bank example, two samples were taken on consecutive Fridays, but the individual subjects were selected at random and there was no relationship between the subjects in one sample and the subjects in the other. The third assumption is that the populations from which the samples are drawn are normally or essentially normally distributed. Note we must say populations at this point because we do not know whether we are working with two populations or one. That is what we are attempting to determine. It turns out that this last assumption, the normality of the underlying population(s), is essential. The test statistic is extremely sensitive to departures from normality, even with large samples. To summarize our assumptions, we have

CONVENTION 7.4.1

We can test for **HOMOGENEITY OF VARIANCE** for two sample variances if:

1. The two samples are independent.
2. The underlying populations are normal or essentially normal.

The British statistician Sir Ronald A. Fisher is credited with the initial work in the area of testing two sample variances. The test actually derives its name from Sir Fisher. It is called the **F-TEST**. Fisher studied the ratio of the variances and found that the sampling distribution of the ratio of variances from two independent samples from populations with equal variances fell into a family of distributions which were dependent upon the two sample sizes.

EXAMPLE 7.4.2: Recall the bank situation where for individual teller waiting lines, we have a sample of size $n_1 = 10$ with a sample variance of $s_1^2 = 4.61$, and for the single waiting line, we have a sample of size $n_2 = 12$ with a sample variance of $s_2^2 = 0.49$.

1. Hypotheses

One claim that the bank manager could make is that the variation in waiting time is reduced for the single waiting line. This claim would translate into the following hypotheses:

H_0: $\sigma_1^2 \leq \sigma_2^2$
H_1: $\sigma_1^2 > \sigma_2^2$

These hypotheses form a one-sided, upper-tailed test.

2. Significance Level
   Choose α=.05

3. Test Statistic
   Select the 2-sample variance F test on the computer. Sometimes this test is done as part of the two-sample independent t-test. We get $F = 9.3975$ and $p = .001$.

4. Conclusion
   Since $p<.05$, we reject the null hypothesis. This means that the data does suggest that waiting time variation has been reduced for the single waiting line.

In this section we are looking at a problem in which we are considering differences in variances between two samples (or populations). We actually began **EXAMPLE 7.4.1** by considering whether or not a single waiting line makes a difference in waiting times from the multiple waiting lines. In this example, the mean waiting times under the two procedures are identical, but there is a noticeable difference between the variances. In the next two sections we will consider problems of difference in means between two samples. We will ask the question, "Does a given difference in sample means suggest that the two samples came from different populations?"

SECTION 5 - DIFFERENCE OF MEANS (INDEPENDENT SAMPLES)

In the previous section we looked at problems where the samples were dependent upon each other. In this section, we will look at cases where the samples are independent. In many such cases the sample sizes will be different.

When we looked at testing hypotheses about means in the previous chapter, we used two test statistics, the z-test and t-test. There are several versions of the 2-sample means test. To decide which one to use, we are going to introduce a series of questions, contained in the following table:

TABLE 7.5.1
TWO-SAMPLE TESTS OF MEANS

Are the two samples dependent?	Are $\sigma 1$ and σ_2 known?	Do n_1 and n_2 BOTH exceed 30?	What conclusion did we draw concerning the hypothesis $s_1^2 = s_2^2$? (F-test)	Test Statistic	EXAMPLE
YES				2-sample dependent t-test	EX. 7.3.1 EX. 7.3.2
NO	YES			2-sample independent z-test	EX. 7.5.1
	NO	YES		2-sample independent z-test	EX. 7.5.2
		NO	Fail to Reject $s_1^2 = s_2^2$ $p \geq \alpha$	2-sample ind. t-test equal variances	EX. 7.5.3
			Reject $s_1^2 = s_2^2$ $p < \alpha$	2-sample ind. t-test unequal variances	EX. 7.5.4

If we determine that our samples are independent (not dependent), the first question we ask is whether the two population standard deviations are known. When would we know population standard deviations? The following example illustrates this type of problem and the methodology we should use.

EXAMPLE 7.5.1: Suppose an admissions counselor at a private college wants to know whether there is a difference between ACT scores for males and females enrolling at her college. It is known that the national standard deviation for men is $\sigma_m = 4.3$ and for women it is $\sigma_f = 4.2$. She selects a sample of 31 male and 31 female ACT scores in the entering class, as shown in the following table:

MALE				FEMALE			
29	18	24	23	20	30	21	21
21	17	26	30	18	21	25	22
25	28	28	19	16	24	24	19
26	21	30	21	19	26	23	27
23	23	24	24	21	19	28	22
32	22	21	26	23	23	20	24
21	18	21	28	25	29	21	20
19	25	20		20	18	25	

This is a two-sample test of difference of means for independent samples with known population standard deviations. The hypothesis testing procedure is as follows:

1. Hypotheses

   $H_0 : \mu_m - \mu_f = 0$ (There is no difference between mean male and female ACT scores)
   $H_1 : \mu_m - \mu_f \neq 0$ (There is a difference between mean male and female ACT scores)

   These hypotheses form a two-sided test, since the admissions counselor is not looking for a particular direction..

2. Significance Level
   We select α=.05.

3. Test Statistic
   We select the 2-sample z-test on the computer and get z=1.16 with p= .06.

4. Conclusion
Since $p>.05$, we retain H_o and conclude there is insufficient evidence to conclude there is a significant difference between the mean composite ACT scores of males and females at this college.

Journal form: $z=1.16, p>.05$.

(Note: Most computers do not have the 2-sample z-test as an option. Instead, a 2-sample t-test would be chosen. In this case, the computer analysis using a 2-sample t-test would yield p=.183. The same conclusion would be reached regarding H_o)

What happens if we do not know the population standard deviations? The next example illustrates what we can do if the sample sizes are large enough. [How large do you think is large enough? What cut-off have we used before?]

EXAMPLE 7.5.2: Suppose the admissions counselor in EXAMPLE 7.4.1 does not know the national [population] standard deviations for men and women. Suppose the only information that she has is the sample deviations, i.e. $s_m = 3.95$ for men, and $s_f = 3.38$ for women. We can follow the same procedure we used in the one-sample case, where we required a large sample of greater than or equal to 30. For the two-sample case, we will need **BOTH** samples to be greater than or equal to 30. If they are, then we can substitute the sample standard deviations for the population standard deviations.

1. Hypotheses

Since the null hypothesis must involve an equality statement, we can use the same hypotheses as EXAMPLE 7.5.1, i.e.,

$H_0 : \mu_m - \mu_f = 0$ (There is no difference between mean male and female ACT scores)
$H_1 : \mu_m - \mu_f \neq 0$ (There is a difference between mean male and female ACT scores)

2. Significance Level
We select $\alpha=.05$.

3. Test Statistic
We select the 2-sample z-test on the computer using the sample means and sample standard deviations and get $z=1.348$ with $p= .044$.

4. Conclusion

Since $p<.05$, we reject H_o and conclude that there is sufficient evidence to suggest there is a difference in the mean composite ACT scores between men and women at this college.

Journal from: $z=1.348$, $p < .05$.

At first it may seem strange that we rejected the H_0 of no difference between male and female ACT scores using the sample variances, but did not reject using the population variances. In both tests, the p values are close to the significance level of .05. In the first example, using the population values, the variances are larger and thus make it more difficult for the difference to be *significantly* different. In the second example, using the sample values, the variances are smaller and thus made it possible to just barely reject the H_o and conclude there is a significant difference between male and female scores.

What happens if one or both of the sample sizes is less than 30? The next question suggested in the introduction to this section was whether the population standard deviations were thought to be the same or different. Thus for problems of this type, we will need to first do an F-test on the sample variances to determine whether it is likely that the two population variances are equal or different. The next example illustrates how we can handle the case when we can't reject the hypothesis that the two variances are equal. The second example illustrates what we can do if the F-test says to reject the hypothesis of equal variances.

EXAMPLE 7.5.3: Suppose an instructor believes that students can learn statistics better if they work in groups similar to the study groups in law school. He encourages his students to set up such study groups. Out of a class of 130 students, about half of the students take his suggestion. To test his belief, the instructor randomly selects 10 students who were involved in study groups and 10 students who chose not to be involved. The instructor is going to test his hypothesis that study groups will improve scores by comparing final averages of the two groups. One student in each of the two groups drops the course before the final exam, and one non-study group student became very ill and had to take an incomplete. Thus, the instructor had usable scores from 8 study group and 9 non-study group students as follows:

Study Group Students	Non-Study Group Students
62	55
97	61
86	76
72	52
91	47
81	56
65	70
77	65
	63

$n_1 = 8$	$n_2 = 9$
$\sum x_1 = 631$	$\sum x_2 = 545$
$\bar{x}_1 = 78.88$	$\bar{x}_2 = 60.55$
$s_1^2 = 151.27$	$s_2^2 = 82.77$
$s_1 = 12.30$	$s_2 = 9.10$

These samples are independent and each is less than 30. Can we assume homogeneity of variances, i.e., can we assume $s_1^2 = s_2^2$? To test variances we need to perform an F-test:

H_0: $\sigma_1^2 = \sigma_2^2$
H_1: $\sigma_1^2 \neq \sigma_2^2$
α=.05
p=.42
Since p>.05, retain H_o and conclude equality of variances

Now we may assume that the variances are equal. What do we do now? We proceed with a test of means.

1. Hypotheses

Since the instructor believes that the Study Group students will perform better than the other students, the question we want to answer is whether m_1 is greater than m_2. Since the null hypothesis must have an equality in it, we have the following hypotheses:

$H_0 : \mu_1 \leq \mu_2$
$H_1 : \mu_1 > \mu_2$, where μ_1 is the mean of the study group and μ_2 is the mean of the non-study group

These hypotheses form a one-sided, upper-tailed test, where rejection of the null hypothesis is the desired result, i.e., the study group students performed better than the non-study group students.

2. Significance level
We choose α=.05

3. Test Statistic
We select the two-sample independent t-test, equal variances, and perform it. We get t=3.52 and p=.0015.

4. Conclusion
Since $p<.05$, reject the null hypothesis of $\mu_1 \leq \mu_2$ and conclude that the evidence suggests that the mean for group 1 (the study group students) is greater than the mean for group 2 (the non-study group students).

Journal form: $t(15) = 3.52, p<.05$.

The final type of problem in this section deals with problems where we do not have homogeneity of variances.

EXAMPLE 7.5.4: Suppose a researcher believes that even small amounts of alcohol will reduce a person's ability to react (increase reaction time). To test her belief, she conducts an experiment in which she measures reaction time of individuals at a party. One group of individuals have not had any alcoholic drinks, while another group of individuals have had one glass of champagne. The researcher measures the response time (in milliseconds) of the individuals to both visual and audible signals. The results of the experiment are given below:

NON-ALCOHOL		ALCOHOL	
21.2	22.3	42.3	49.6
25.3	29.6	50.2	35.2
14.5	24.5	45.3	15.2
18.3	21.2	63.5	38.2
15.2	20.3	16.3	12.5
19.6	13.2	19.2	71.2
26.3	23.3	44.9	18.7
30.5	15.9	17.3	
21.1		32.3	

Non-Alcohol Group	**Alcohol Group**
$n_1 = 31$	$n_2 = 16$
$\bar{x}_1 = 19.60$	$\bar{x}_2 = 27.20$
$s_1^2 = 43.82$	$s_2^2 = 197.96$
$s_1 = 6.62$	$s_2 = 14.07$

Do these results collaborate the researcher's belief?

This is a two-sample test of means with one small sample. Since one of the sample sizes is less than 30, we need to determine whether homogeneity of variance is a tenable assumption. For these two samples, the F test gives us $p < .0001$. If we assume a level of significance of $\alpha = 0.05$, $p<.05$. Thus, we must reject the hypothesis of homogeneity of variance. This means that we must assume that $\sigma_1^2 \neq \sigma_2^2$. Now we perform the test for means:.

1. Hypotheses

   The researcher believes that the non-alcohol group has faster response time than the alcohol group. Her research hypothesis is that $\mu_1 < \mu_2$, where μ_1 is the mean of the non-alcohol group. This translates into the following set of hypotheses:

   $H_0 : \mu_1 \geq \mu_2$
   $H_1 : \mu_1 < \mu_2$

   This is a one-sided, lower-tail test for which the researcher hopes the calculated value does fall within the rejection region.

2. Significance Level
   Choose $\alpha=.05$

3. Test Statistic
   Select the 2-sample independent t-test, unequal variances, on the computer. We get $p=.003$.

4. Conclusion
   Since $p<.05$, we may reject the null hypothesis. Therefore, the evidence suggests that the non-alcohol group does indeed have better response (shorter response time).

   Journal form: $t(45) = -2.049$, $p < .05$.

SECTION 5 - SUMMARY

In this chapter we studied two-sample tests. We first considered tests of difference of means for dependent samples. The experimental design for many such tests will either be the matched pair design or the before-and-after design.

We then looked at two-sample tests of difference of means for independent samples. There were four possible tests depending upon the size of the samples, whether the population standard deviations were known, and whether we could assume homogeneity of variance. In order to choose the appropriate form of the independent test, we learned how to perform the F-test, a test of homogeneity of variance.

KEY WORDS AND CONCEPTS

Homogeneity of Variance
F-test
Before-and-After Experimental Design
Independent Samples
Dependent Samples
Matched-Pair Design

KEY FORMULAS

Difference of Means (dependent samples)

$$t = \frac{\bar{d} - \mu_d}{\frac{s_d}{\sqrt{n}}}$$

Difference of Means (independent samples, known $\sigma's$)

$$z = \frac{(\bar{x}_1 - \bar{x}_2) - (\mu_1 - \mu_2)}{\sqrt{\frac{\sigma_1^2}{n_1} + \frac{\sigma_2^2}{n_2}}}$$

Difference of Means (independent, large samples, unknown $\sigma's$)

$$z = \frac{(\bar{x}_1 - \bar{x}_2) - (\mu_1 - \mu_2)}{\sqrt{\frac{s_1^2}{n_1} + \frac{s_2^2}{n_2}}}$$

Difference of Means (independent, small samples, homogeneous variances)

$$z = \frac{(\bar{x}_1 - \bar{x}_2) - (\mu_1 - \mu_2)}{\sqrt{s^2\left(\frac{1}{n_1} + \frac{1}{n_2}\right)}}$$

where $s^2 = \frac{(n_1 - 1)s_1^2 + (n_2 - 1)s_2^2}{n_1 + n_2 - 2}$ and $df = n_1 + n_2 - 2$

Difference of Means (independent, small samples, nonhomogeneous variances)

$$t = \frac{(\bar{x}_1 - \bar{x}_2) - (\mu_1 - \mu_2)}{\sqrt{\frac{s_1^2}{n_1} + \frac{s_2^2}{n_2}}}$$

CHAPTER 8
CATEGORICAL DATA

SECTION 1 - INTRODUCTION

We begin this chapter by referring to the soap problem posed in **A Word to the Student**. Recall that your company is involved in the manufacture of facial soap. The natural color of the soap is *muddy brown*. Is muddy brown a good color for soap? Does the color of soap make a difference to consumers? Should you introduce an additional step in the process, with associated costs, to change the color to something *more attractive* to your potential customers? You could run a marketing test to determine if there are large enough differences in customer preference so that it will pay to introduce a different color to the soap. You provide 100 homes with samples of three *different* facial soaps which you are testing. Although you imply that there are *significant differences* in the soaps, the only real difference is the color. You leave one the original muddy brown; you color one white and the other pastel pink. You ask the families to use the three soaps for several weeks. After four weeks you call the families and ask them how the test is progressing. You ask them which of the soaps they prefer and why. The families report the following preferences:

Soap	Muddy Brown	White	Pink
Number of families	10	60	30

White and pink are preferred much more than the natural color. Are the differences enough to suggest that you do color the soap? This type of problem is called a **categorical problem**. We are dividing data into different categories. This particular problem is called a **one-dimensional categorical problem**. It involves just one variable. Problems of this type can involve frequencies (as the one above) or proportions. In Section 2 of this chapter we introduce a statistical procedure in which we test **observed** frequencies against **expected** frequencies.

In the problem above, if color *made no difference*, then we would expect the frequencies to be about the same. This type of test is called a **goodness-of-fit test**. With this test, we are testing how well our observations conform (*fit*) to our expectations. The test statistic used in a goodness-of-fit test is known as the **χ^2 (chi-square)** statistic. We will introduce this family of distributions in Section 2.

In Section 3, we turn to **two-dimensional categorical problems**. We analyze tables of data in which the rows represent categories for one variable and the columns categories for another variable. Questions like "Are smoking and lung cancer linked?" or "Does socio-economic status influence opinions concerning gun control?" are examples of two-dimensional categorical problems. The frequencies can be reported in two-dimensional tables which we call **contingency tables**. In the two questions cited above, we are interested in whether there is a connection between two variables. We are asking if the two variables are **independent** or **dependent**. We will use the χ^2 test to determine independence.

For both χ^2 tests, the data is usually collected and put into the computer as raw data. Some computer programs require that it first be put in tabular form before the test can be performed; others will automatically perform the test on the raw data.

SECTION 2 - GOODNESS-OF-FIT TEST

In this section, we study a situation in which the outcomes belong to one, and only one, of several categories and determine whether the data "fits" some hypothesized distribution.

EXAMPLE 8.2.1 Let's return to the soap problem from Section 1. We have the following table, with the data already summarized:

TABLE 8.2.1

Soap	Muddy Brown	White	Pink
No. of Families	10	60	30

If there were no difference in color preference, we would expect that there would be approximately the same number of families selecting each of the three colors. There would be 33.33 families selecting each color, and we would have the following table of both observed and expected frequencies:

TABLE 8.2.2

Soap	Muddy Brown	White	Pink
Observed No. of Families	10	60	30
Expected No. of Families	33.33	33.33	33.33

We can see a difference between the observed values and the expected values. Are these values *different enough* to be significant? Our hypotheses would look as follows:

H_0: uniform distribution (same expected frequencies for each category)
H_1: not a uniform distribution

To test whether there are significant differences between the observed and expected frequencies, we first introduce a measure of *disagreement* between them. It is important to note that we are using frequencies in this test. If the original problem is stated in terms of proportions, we need to convert the proportions to frequencies. [How?] In this measure of disagreement, we use the following notation:

NOTATION 8.2.1	
O	represents the **observed frequency** of an outcome.
E	represents the **expected frequency** of an outcome.
k	represents the **number of categories** or outcomes.
n	represents the **total frequency or sample size**.

From TABLE 8.2.2, we see that the values of **O** for this problem are 10, 60 and 30. The values for **E** are 33.33, 33.33 and 33.33. Note that Σ**O** = Σ**E** = 100. Since there are 3 categories, the value for **k** is 3. How do we proceed to measure the disagreement between observed and expected frequencies? The χ^2 Goodness-of-Fit Test calculates a χ^2 test statistic which determines whether the total accumulated discrepancy between the **O**s and **E**s is large enough to be "significant," i.e. cannot be attributed to chance alone. The computer uses the formula

$$\chi^2 = \sum \frac{(O-E)^2}{E}$$

The symbol χ^2 (chi-square) represents a family of distributions. There is a different distribution for each number of categories. In addition, the family of distributions has the following properties:

1. The values of chi-square can be zero or positive, but they can never be negative.
2. The chi-square distributions are not symmetric, but are skewed to the right.

The computer evaluates the χ^2 for a problem and gives us its corresponding p value. If $p>\alpha$, we retain H_o and if $p<\alpha$, we reject H_o.

If we return to the soap problem, the computer gives us a calculated value of χ^2=38, or a p value of 5.59×10^{-9}. If we set our significance level at 5%, we would reject H_0 since

$5.59 \times 10^{-9} < .05$. We would reject a null hypothesis that the observed frequencies are equal to the expected frequencies. In this case, the color of the soap does make a difference.

In this first example, all of the expected frequencies are equal. Consider the following example where the expected frequencies are not the same:

EXAMPLE 8.2.2: Suppose you are the personnel director of a small manufacturing plant. You are concerned about employee absences for illness, particularly on Mondays and Fridays. [Why Mondays and Fridays?] Historically, employee absences have followed the following pattern:

TABLE 8.2.3

Days of the Week	Monday	Tuesday	Wednesday	Thursday	Friday
Percentage of Reported Absences	37.5%	12.2%	10.7%	9.8%	29.8%

In order to decrease the absences before and after a weekend, you institute a policy of "docking" workers an extra day's pay for an unexcused absence. An absence is *unexcused* if the employee cannot produce a doctor's note within two working days of a return to work. During the first three-month period of the new policy, there were 83 days of employee absence. The actual days of absences are listed in the following table:

TABLE 8.2.4

Day of the Week	Monday	Tuesday	Wednesday	Thursday	Friday
Observed Absences	21	16	14	13	19

If the new policy had no effect on employee absences, how many absences would we expect, from our historical records, for each day of the week? On Mondays, historically there were 37.5% absences. Therefore, we should expect 37.5% of the 83 absences on Mondays. (83 times 0.375 is 31.125) If we do this for each day, we have the following table:

TABLE 8.2.5

Day of the Week	Monday	Tuesday	Wednesday	Thursday	Friday
Observed Absences	21	16	14	13	19
Expected Absences	31.13	10.13	8.88	8.13	24.73

1. Hypotheses

H_0: population distributed according to the historical pattern
H_1: population not distributed according to the historical pattern

You want to see if there is a difference between the historical pattern and the new pattern. Has the new policy made a difference?

2. Significance level
Choose a = .05

3. Test Statistic
Choose the Chi-Square Goodness-of-Fit Test. From the computer, χ^2=13.8949 or p=.0076.

4. Conclusion
Since our $p < .05$, we must reject the null hypothesis that the observed frequencies equal the expected. In this case, it seems that the new policy has had an effect on absences.

Journal form: χ^2 (4) = 13.8949, $p < .05$.

The two examples above illustrate what we do with one-dimensional categorical or goodness-of-fit problems. The expected frequencies will be determined either by historical records or by assuming the expected frequencies satisfy some probability distribution.

SECTION 3 - TEST OF INDEPENDENCE

In the previous section, we looked at one-dimensional categorical problems. In this section, we look at what we will call two-dimensional categorical problems. Consider the following example:

EXAMPLE 8.3.1: Suppose we were interested in whether the color of a person's hair had any direct correspondence to the color of their eyes. We sample 100 individuals and find the following data:

TABLE 8.3.1

Observed Values		**Color of Hair**		
		Brown	Blonde	*Marginal Row Totals*
Color of Eyes	Brown	47	15	*62*
	Blue	9	29	*38*
Marginal Column Totals		*56*	*44*	*100*

We see that 62 people had brown eyes, while 38 had blue eyes. We see that 56 people had brown hair, while 44 had blonde hair. Is there some relationship between the color of eyes and the color of hair? We have observations in Table 8.3.1, sometimes called a contingency or pivot table. What values could or should we expect?

If there were no relationship between the color of hair and the color of eyes, would we expect the same number of people in each of the four blocks? That is one way to look at the problem. However, we know that there are not equal numbers of brown and blonde-haired people. We need to somehow take that into account.

From the marginal totals, we know that 56% of the sample had brown hair. We could assume that 56% of the brown-eyed people also had brown hair. If there are 62 brown-eyed people, then 56% of 62 is 34.72.

What would happen if we looked at the situation in the other order? From the marginal totals, we know that 62% of the people had brown eyes. Therefore, we could expect 62% of the brown-haired people to have brown eyes. There are 56 brown-haired people. Thus, we could expect 62% of 56, or 34.72 people to have both brown hair and brown eyes.

We could do this for each of the entries in the table. We are multiplying the appropriate marginal row total times the appropriate marginal column total and dividing by the total sample size. If we did this for Table 8.3.1, we would have the following expected values:

TABLE 8.3.2

Expected Values		**Color of Hair**		
		Brown	Blonde	*Marginal Row Totals*
Color of Eyes	Brown	34.72	27.28	*62*
	Blue	*21.28*	*16.72*	*38*
Marginal Column Totals		*56*	*44*	*100*

In other words, for each cell: $E = \frac{(row\ total)(column\ total)}{grand\ total}$

Note in this table all of the marginal totals are exactly the same as the marginal totals for the observed values. [Why?] Instead of creating a completely separate table, in practice we will set up a "little corner of each cell" for the expected values. It will look like the following:

TABLE 8.3.3

Observed and Expected Values		**Color of Hair**		
		Brown	Blonde	*Marginal Row Totals*
Color of Eyes	Brown	47 / 34.72	15 / 27.28	*62*
	Blue	9 / *21.28*	29 / 16.72	*38*
Marginal Column Totals		*56*	*44*	*100*

We use the same formula to calculate the χ^2 test statistic, taking the difference of the observed and expected frequencies. Using a computer, we get χ^2=25.977 or p=3.57x10^{-7}. At the 5% level of significance, since p<.05, we would reject the hypothesis that the observed frequencies equal the expected. Therefore, the color of a person's hair and the color of that person's eyes are related. The two variables (color of hair and color of eyes) are **DEPENDENT**. They are not **INDEPENDENT**.

Problems of this type are sometimes called **TESTS OF INDEPENDENCE**. We are trying to determine whether two variables are related (dependent upon each other). Tests of this type do not tell us what type of relationship exist, only if a relationship is likely. The χ^2 Test of Independence uses the following steps:

1. Hypotheses
   H_0: the two variables are independent (no relationship between them)
   H_1: the two variables are dependent (relationship between them)

2. Significance Level
   Choose α

3. Test Statistic
   Select the χ^2 Test of Independence on the computer and get the p value

4. Conclusion
   If $p>\alpha$, retain H_o and conclude the variables are independent
   If $p<\alpha$, reject H_o and conclude the variables are dependent

The previous problem had the same number of categories for each of the two variables. The following example illustrates a situation when there are not the same number of categories for the two variables.

EXAMPLE 8.3.2: Suppose you represent a potato chip company. Your company can use any one of three types of packaging: foil, waxed paper or cellophane. Cellophane is the cheapest; foil is the most expensive. Because of the cost of the wrapping paper, you might be tempted to use only cellophane. However, you know that some people prefer the other two kinds of paper and, in fact, will not buy chips in cellophane bags. The usual rationale given by consumers is that cellophane is easily torn and will not block out sunlight. Both of these situations tend to accelerate the spoilage rate of chips.

Your experience in market research tells you that the buying patterns of consumers are different in different geographic sectors. You decide to do a market study to see if you need to use different packaging policies for different types of communities. You bag chips in the three different packages of the same size, at the same price [Why?] You select test stores in four different types of communities: inner city; suburbs; medium size city; and a small, rural town. After a two-week test, you have the following results:

TABLE 8.3.4

Observed Values		Type of Package			*Marginal Row Totals*
		Cellophane	Waxed	Foil	
Test Markets	Inner City	15	35	54	*104*
	Suburb	63	11	34	*108*
	Medium City	57	32	23	*112*
	Rural Town	12	48	16	*76*
Marginal Column Totals		*147*	*126*	*127*	*400*

The table indicates that 400 packages of chips were sold during the one-week trial. We need to determine the expected values, by multiplying marginal row and column totals and dividing that product by 400. For example, the expected value of cellophane packaging in the inner city is

$$E = \frac{(104)(147)}{400} = \frac{15288}{400} = 38.22$$

The computer would generally do this and the resultant table is as follows:

TABLE 8.3.5

Observed and Expected Values		Type of Package			*Marginal Row Totals*
		Cellophane	Waxed	Foil	
Test Markets	Inner City	15 38.22	35 32.76	54 33.02	*104*
	Suburb	63 39.69	11 34.02	34 34.29	*108*
	Medium City	57 41.16	32 35.28	23 35.56	*112*
	Rural Town	12 27.93	48 23.94	16 24.13	*76*
Marginal Column Totals		*147*	*126*	*127*	*400*

From the computer, $\chi^2 = 103.702$ or $p = 4.23 \times 10^{-20}$. For $\alpha = 0.05$, since $p<.05$, we would reject the null hypothesis that the variables are independent (observed equals expected). The evidence suggests that there *is* a relationship between the geographic area and packaging preference. The Chi-Square test, however, does not indicate *how* the two variables are related. You would have to do further investigation to determine that.

SECTION 4 - SUMMARY

In this chapter we looked at categorical problems. One-dimensional categorical problems involve one variable which has a number of possible outcomes. We looked at how close observed values match expected values. This type of problem is called a χ^2 Goodness-of-Fit Test. In two-variable categorical problems, we investigated whether there is a relationship between the two variables. This type of test is called a χ^2 Test of Independence.

KEY WORDS AND CONCEPTS

Categorical Problem
Observed Frequencies
Expected Frequencies
Contingency Table
Goodness-of-Fit Test
Chi-Square Distribution
Test of Independence

KEY FORMULAS

Chi-Square $\qquad \chi^2 = \sum \frac{(O-E)^2}{E}$

χ^2 Goodness-of-Fit degrees of freedom $\qquad$ #categories - 1

χ^2 Test of Independence degrees of freedom $\qquad$ (#rows - 1)(#columns - 1)

χ^2 Test of Independence expected value $\qquad E = \frac{(row\ \ total)(column\ \ total)}{grand\ \ total}$

CHAPTER 9
CORRELATION ANALYSIS

SECTION 1 - INTRODUCTION

In the previous several chapters, we constructed hypotheses and performed hypothesis tests. In this chapter, we introduce another type of hypothesis test. But we first want to return to some of the concepts of descriptive statistics to look at the idea of relationship between two variables. One of the most important concepts to learn in all of statistics (in all of mathematics, and in almost all other disciplines) is that the relationship between two variables drives the way you look at and analyze the data. Consider the following data:

TABLE 9.1.1

	A	B	C	D	E	F	G	H
x	67	73	87	47	52	68	59	83
y	78	83	91	69	73	81	68	87

Now suppose columns A through H represent students in a mathematics course. Row x represents the students' scores on a pre-test, and row y represents the same students' scores on a post-test. In this situation, each student has two scores. The scores are related by virtue of the fact that they are the scores of a given individual. What would we do with this problem? We could ask whether the course helped the students do better on the test. To see if they did do better, we could treat this problem as a "Before-and-After Experimental Design" and perform a dependent, two-sample difference of means test.

However, we could take this same data and look at it in a different light. We could ask the question, "What is the relationship between the pre-test score and the post-test score?" Could we "predict" what a student will do on the post-test, if we knew what he or she did on the pre-test? We are asking whether there is a "mathematical" relationship between the pre-test score and the post-test score. If there were a relationship which can be represented by a mathematical formula, we could then use that relationship to predict the score on the post-test from the individual's score on the pre-test. We are using the same data, but we are asking two different questions about that data.

In this chapter we will develop the concept of a mathematical relationship between two or more variables. In Section 2 we will construct a type of graph, called a **scattergram**, which will be used to picture the data to visualize the type and strength of any relationship that may exist. We will also develop the idea of a linear relationship. We will develop a measure of linearity, i.e., a measure of how close to a straight line the data falls. This measure will be called the **correlation coefficient**, and the "best fit" straight line will be called the **regression line**. In the next section we will develop a test of significance for the correlation coefficient. We will use this to determine when we can use the regression line to make predictions. In Section 4, we will develop the prediction formula, or equation of the regression line.

If the regression line developed in Section 4 is not an "exact" fit (and it won't be, unless the original data was a straight line), then there will be differences or **variations** between the data and the predicted values. In Section 5 we will study this variation.

SECTION 2 - CORRELATION

In the previous section, we noted that we will begin by plotting data to **SEE** if there is any relationship between the variables. Visualization of the data often helps us to determine an intuitive feel for the data, and gives us insight into possible mathematical relationships. The following diagram represents a plot of the data from TABLE 9.1.1. We have selected one variable to plot along the x-axis, which we will call the **independent variable**, and one variable along the y-axis, which we will call the **dependent variable**.

Figure 9.2.1

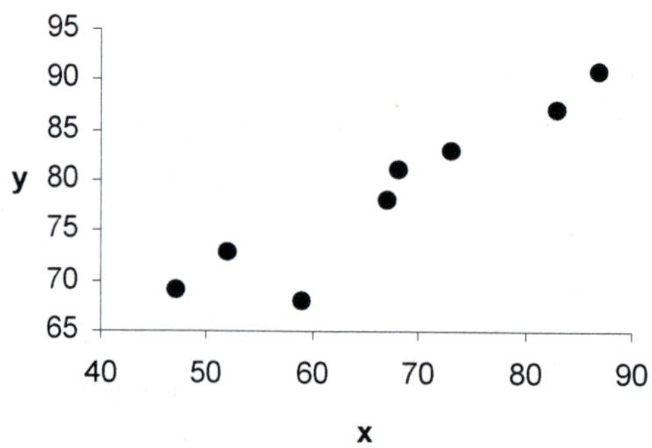

This data does not fall on a perfect straight line. However, the data is close to a straight line. From this graph we would intuitively conclude that there is a relationship between the x and y variables, and that relationship is close to being linear. The graph above is called a **SCATTERGRAM** or **SCATTER DIAGRAM**.

The following scattergrams represents several different types of relationships.

Figure 9.2.2 (a)

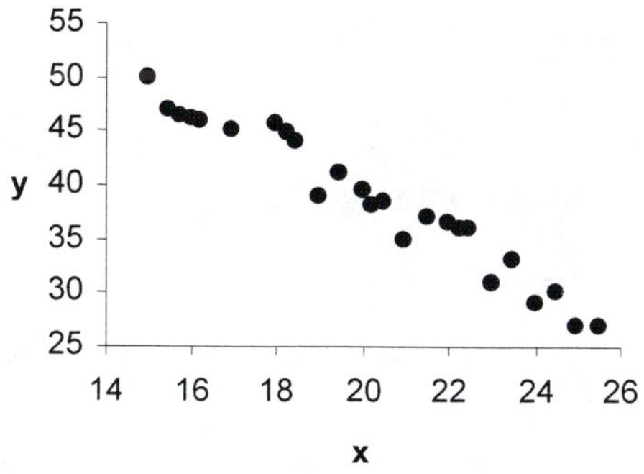

Figure 9.2.2 (b)

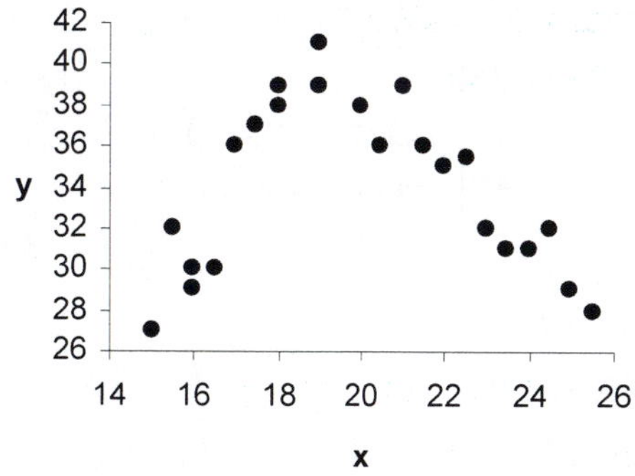

Figure 9.2.2 (c)

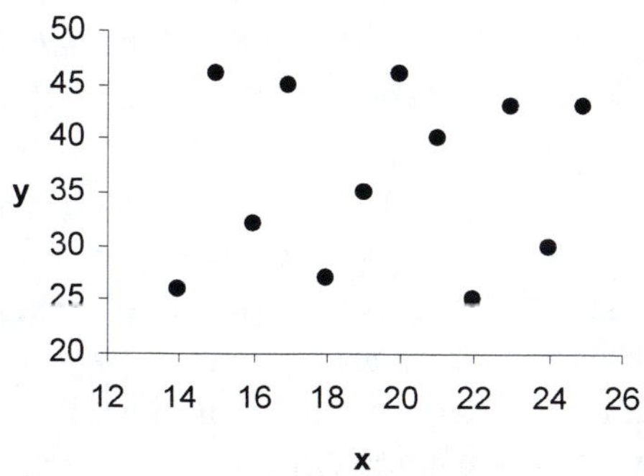

Figure 9.2.2 (d)

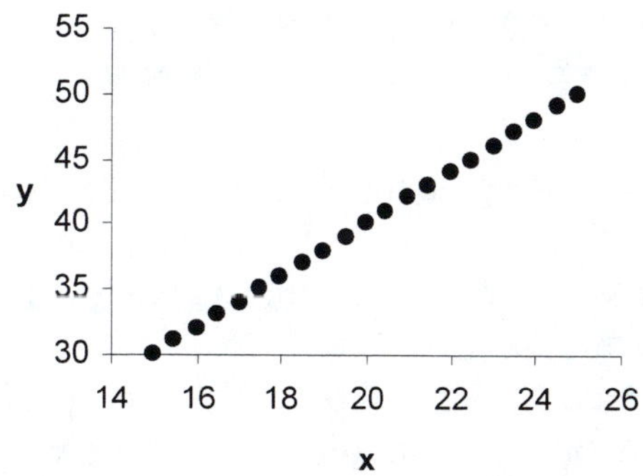

In Figure 9.2.1, as x (the independent variable) increases, the values of y (the dependent variable) tend to increase also. We say that this is a **positive** relationship. In Figure 9.2.2 (a), as the values of x increase, the values of y tend to decrease. We say that this is a **negative** relationship. Figure 9.2.1 does show something of a linear relationship, while the data in Figure 9.2.2 (a) fall closer to a straight line. We will say that the linear relationship in Figure 9.2.2 (a) is **stronger** than the relationship in Figure 9.2.1.

The data in Figures 9.2.2 (b) and (c) are not linearly related. In Figure 9.2.2 (c), there is no apparent (or at least not readily apparent) relationship. In Figure 9.2.2 (b), the data is related, but that relationship is not linear. In this case, it appears to be a quadratic relationship. Figure 9.2.2 (d) is a perfect straight line. The above discussion leads us to the following definition:

> **DEFINITION 9.2.1**
> **CORRELATION** is the degree of closeness to a linear relationship between two variables. The closer to a straight line, the **STRONGER** the correlation.

Thus, we can say that the data in 9.2.1 is positively correlated; the data in 9.2.2 (a) is strongly, negatively correlated; and the data in 9.2.2 (d) is perfectly positively correlated. The data in 9.2.2 (b) and (c) are not correlated (or at most very weakly correlated). **REMEMBER:** Data may be related (such as a quadratic relationship), but not linearly correlated.

In the definition and discussion we talked about correlation being strong or weak. How do we measure it? For the figures presented, we can see the difference. It will not always be possible to see "differences." We need something more definite. To do this we introduce the concept of what we will call the **CORRELATION COEFFICIENT**. This mathematical entity was first discovered by the statistician Karl Pearson (1857-1936). Sometimes it is called **PEARSON'S PRODUCT MOMENT**, **PEARSON'S CORRELATION COEFFICIENT** or **PEARSON'S r.**

Before we give the formula for Pearson's Correlation Coefficient, we should discuss several properties of Pearson's r. First, we note that correlation can be positive or negative. Positive correlation means that as the independent variable increases, the dependent variable also increases. Negative correlative means that as the independent variable increases, the dependent variable decreases. Pearson's r will be positive for positive correlation, and negative for negative correlation.

Pearson's r is formulated in such a way that a perfect straight line will have an r-value of +1 or -1, depending upon whether the straight line has positive or negative slope. The closer to a straight line the data falls, the closer to $\pm$ 1 is the Pearson's r. For values of Pearson's r which are close to 0, the data is far from a linear relationship. But we need to remember that this does not necessary mean the data are not related. They are not related linearly. In summary,

The **PEARSON r** is a measure of “how close to a straight line” data falls. It has the following properties:

(1) $-1 \leq r \leq +1$

(2) $r > 0$ indicates a relationship in which the dependent variable increases as the independent variable increases

(3) $r < 0$ indicates a relationship in which the dependent variable decreases as the independent variable increases

(4) $r = \pm 1$ indicates the data falls in a "perfect" straight line

(5) $r = 0$ indicates that there is no linear relationship between the variables

If we were to compute Pearson's r by hand, we would use the following formula (only one of several equivalent formulas):

$$r = \frac{n\Sigma xy - \Sigma x\Sigma y}{(\sqrt{n\Sigma x^2 - (\Sigma x)^2})(\sqrt{n\Sigma y^2 - (\Sigma y)^2})}$$

where:

n	is the number of pairs of data points
Σxy	is the sum of the products of "x times y"
Σx^2	is the sum of the squares of the x's
$(\Sigma x)^2$	is the square of the sum of the x's

Computers, however, can compute correlations with great ease!

EXAMPLE 9.2.1 Find the Pearson's r for the data from TABLE 9.1.1.

Using a computer, $r = 0.942$. Since the graph of the data is very close to a straight line, a correlation so close to 1 is not surprising. In other words, there is a strong, positive linear association between students' scores on a pretest and a posttest.

Having calculated the Pearson's r, what do we do with it? In the example above, the Pearson's r is close to 1. Therefore, the data is close to a straight line. There are two questions which we ask. The most obvious question is "What straight line?". That question will be answered in Section 4 of this chapter. A question which is not as obvious, but must come first, is the question, "Is the correlation **really** significant?" Is it close enough to 1 to use the regression line as a predictive tool? The significance of correlation is treated in Section 3.

SECTION 3 - TEST OF SIGNIFICANCE OF PEARSON CORRELATION COEFFICIENT

In the previous section, we calculated the Pearson's r for a set of data. We indicated that the next question is whether that r-value is significant. We need to do a hypothesis test for significance. What are we testing?

The data with which we are working is most likely a sample (or at least treated like a sample). Thus, the calculated r-value is really a sample statistic which estimates a population parameter. What we would like to know is that population parameter. Thus, we can hypothesize a value for that population parameter, and test it against the calculated value.

Usually we hypothesize in the null hypothesis that there is no correlation and in the research hypothesis that there is a significant correlation:

H_0: $\rho = 0$
H_1: $\rho \neq 0$ (or $\rho > o$, $\rho < 0$) where ρ is the population correlation

EXAMPLE 9.3.1: If we use the correlation that we calculated in EXAMPLE 9.2.1, we have $n = 8$ and $r = 0.942$. We test this correlation against a hypothesis of $\rho = 0$ using α=.05.

1. Hypotheses
   $H_0 : \rho = 0$
   $H_1 : \rho \neq 0$

2. Significance level
   Choose α=.05

3. Test Statistic
   From the computer, we get r=.942 and p=.0005

4. Conclusion
   Since p<.05, we reject H_0 and conclude there is significant correlation between pre and post scores.

What do we do now? If we have shown significant correlation, then we can proceed to find the regression line and use it to make predictions about other data points. It is extremely important to remember that we **should not use a regression line to predict values unless the correlation is significant**. Even if the correlation is significant, we need to also remember that the use of a regression line will not provide completely accurate estimates or predictions. There will be errors or variation. The regression line is the line which minimizes those errors.

SECTION 4 - REGRESSION LINE

In the previous sections we calculated the correlation coefficient, a measure of linearity of our data, and then tested the statistical significance of that correlation. In this section, given data with significant correlation, we will find the "best fit" straight line, or **REGRESSION LINE**, for our data. We first need to explain what we mean by "best fit." Suppose we look at the data with which we have been working in this chapter. Recall that the table of values and scattergram look like the following:

TABLE 9.4.1

	A	B	C	D	E	F	G	H
x	67	73	87	47	52	68	59	83
y	78	83	91	69	73	81	68	87

Figure 9.4.1

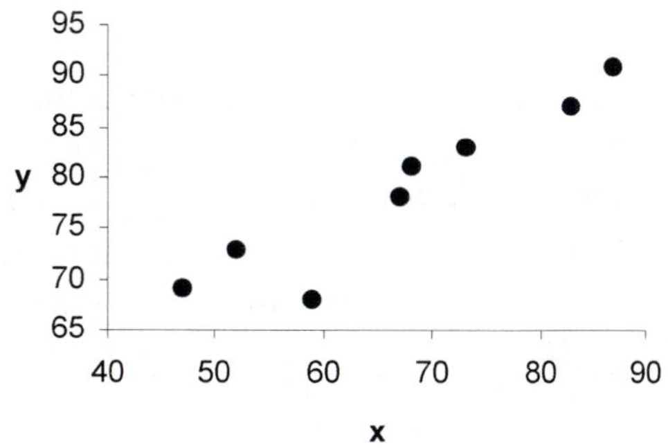

What do we mean by a "best fit" straight line? Since the data does not lie on a straight line, for many, if not all data points, there will be variations or errors between the data points and the line. What do we do with all those variations. We are going to do something similar to what we did when we introduced the concepts of variance and standard deviation. We measure the deviation of the data points from the line. Just as the standard deviation minimized the squared deviations from the mean, the regression line will be the line which minimizes the squared deviations from the data points. We have the following definition.

> **DEFINITION 9.4.1**
> The **REGRESSION LINE** for a set of data is the line which minimizes the squared deviations from the line.

Having defined the regression line, how do we determine its equation? The equation of a straight line takes the form

$$y = A + Bx$$

where

y	is the dependent variable
x	is the independent variable
B	is the slope of the line
A	is the y-intercept of the line

The regression line is completely determined if we can determine its slope B and y-intercept A. By means of the calculus, we can determine the following formulas:

$$B = \frac{n\sum xy - \sum x \sum y}{n\sum x^2 - (\sum x)^2}$$

$$A = \bar{y} - B\bar{x}$$

Usually, however, we let the computer perform these calculations for us!

EXAMPLE 9.4.1: Consider the data in TABLE 9.4.1. Using hand calculations we would get B=.559 and A=41.30 to give us the regression equation y=41.30+.559x.
Using the computer, we would get a table which would give us a constant (means the same as intercept) and coefficient (means the same as slope). We would get the same regression equation y=41.4+.559x.

We can plot this line on the scattergram as follows:

Figure 9.4.2

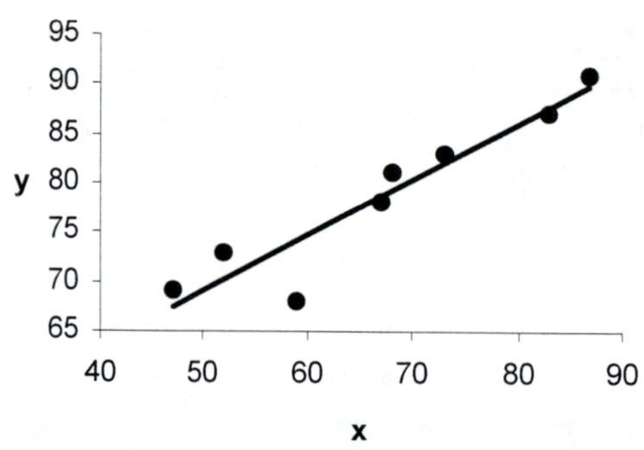

In the examples above, we calculated the Pearson's r correlation coefficient for a set of data, showed that the r is significant (at the 0.05 level), and determined the regression equation. What can we do with the regression equation? Consider the following example.

EXAMPLE 9.4.2: Consider the situation described in the previous examples, i.e., a pre-test and post-test given in a mathematics course. The correlation coefficient for the data was found to be 0.942, which is significant at the 0.05-level. The regression equation was determined to be

$$y = 41.3 + .559x$$

What would the predicted score be for a student who scores 58 on the pre-test? We can substitute the x-value of 58 into the regression equation and get

$$y = 41.3 + (.559)(58) = 73.72$$

Thus, a student who scores a 58 on the pre-test can be expected to score approximately 74 on the post-test.

In making predictions from regression equations, there are several cautions which we must observe. Disregarding these dangers can have disastrous consequences for the careless researcher. Beware of:

1. **Using the regression equation to make predictions for data for which the correlation coefficient is not significant**. If the correlation coefficient for a set of data is not significant, the best estimate of the dependent variable for any value of the dependent variable is the mean of y.

2. **Using the regression equation to make predictions for values which are outside the range of the original data**. If we try to extend the range of the predictions, we may be in trouble. Our original data may have placed us in a range in which we were close to a straight line. Outside that range, the relationship may change and no longer be a straight line.

3. **Using the regression equation to make predictions for one set of data based upon another set of data**. We need to be careful about "transferring" data. The "new" data may not have the same relationship as the "old" data had. This is particularly true for situations in which our original data is several years old. A new population may have changed considerably.

SECTION 5 - VARIATION

In the preceding three sections we talked about correlation and regression. The correlation coefficient can be used to measure how closely the given data fits a straight line. In looking at correlation coefficients in this way, we are really limited to three conclusion:

1. There is a significant positive linear correlation.
2. There is a significant negative linear correlation.
3. There is not a significant linear correlation.

It is possible to get more information from r. Consider the following example:

EXAMPLE 9.5.1: Suppose we have the following set of pairs of data points:

	1	2	3	4	5	6
y	4	6	10	13	21	18

If we were to calculate the correlation coefficient r and the regression equation $y = A + Bx$, we would get the following:

$$r = 0.947$$

$$y = 0.2 + 3.371x$$

The correlation r=.947 cannot be interpreted to be a percentage. In other words, this correlation is not 94.7% of the way between 0 (no correlation) and +1 (perfect positive correlation). But r can be squared, forming the **coefficient of determination** r^2, which is technically the ratio of the explained variation and the total variation. What does the coefficient of determination do for us? It tells us the percent of variability in the dependent variable that can be accounted for or explained by the independent variable.

For this example, $r^2=(.947)^2=.8968$. Thus, 89.68% of the variability in the dependent variable y can be accounted for or explained by the variability in the independent variable x. The other 10.32% of the variability in y is unexplained, i.e. is explained by variables other than x.

SECTION 6 - SUMMARY

In this chapter, we looked at the relationship between independent and dependent variables. We drew scattergrams. We measured how close to a straight line the relationship was by means of the correlation coefficient. We determined when a correlation coefficient was significant for a set of data. When a correlation was significant, we determined the best fit straight line, by means of the regression equation. Finally, we looked at variation and measured it by the coefficient of determination.

KEY WORDS AND CONCEPTS

Independent variable
Dependent variable
Scattergram
Correlation
Variation
Correlation coefficient
Regression line
Coefficient of determination

KEY FORMULAS

Correlation Coefficient:

$$r = \frac{n\sum xy - \sum x \sum y}{(\sqrt{n\sum x^2 - (\sum x)^2})(\sqrt{n\sum y^2 - (\sum y)^2})}$$

Regression Line

$$y = A + Bx$$

where

$$B = \frac{n\Sigma xy - \Sigma x \Sigma y}{n\Sigma x^2 - (\Sigma x)^2}$$

and

$$A = \overline{y} - B\overline{x}$$

Coefficient of Determination r^2

APPENDIX 1 - DATA SET

VITAL STATISTICS FOR 1987 HIGH SCHOOL SOCCER TEAM

NUMBER	POSITION	GRADE	HEIGHT	WEIGHT
27	D	12	5'11"	160
28	S	12	6'	155
20	M	12	5'9"	147
6	D	12	6'	200
10	S	12	6'	200
30	S	12	5'4"	148
1	G	12	6'3"	150
15	D	12	5'11"	155
25	D	12	5'10"	158
29	M	12	5'9"	135
22	S	12	6'	145
17	S	11	5'8"	135
5	D	11	5'5"	130
9	D	11	6'	145
18	D	11	6'	140
4	S	11	5'11"	155
16	M	11	5'7"	131
1	G	11	5'10"	155
7	D	11	5'9"	140
13	S	11	5'6"	120
31	M	11	5'9"	140
3	S	10	5'8"	125
14	S	10	5'8"	130
32	M	10	6'	140
11	M	10	5'10"	160
34	M	10	5'9"	145
2	S	10	5'4"	105
24	M	10	5'6"	130
1	G	10	5'8"	130
23	S	10	5'5"	120
12	S	9	5'4"	102
19	D	9	5'4"	110
35	S	9	5'11"	140
38	S	9	5'4"	98
39	S	9	5'3"	95
26	M	9	5'5"	105
36	D	9	5'11"	140
8	M	9	5'6"	125
37	G	9	6'	160
21	D	9	5'8"	133
33	M	9	5'2"	116
1	G	9	5'9"	133
40	S	9	5'6"	135
41	M	9	5'10"	120

APPENDIX 2

NORMAL PROBABILITY DISTRIBUTION

Second Decimal

z	0.00	0.01	0.02	0.03	0.04	0.05	0.06	0.07	0.08	0.09
0.0	0.0000	0.0040	0.0080	0.0120	0.0160	0.0199	0.0239	0.0279	0.0319	0.0359
0.1	0.0398	0.0438	0.0478	0.0517	0.0557	0.0596	0.0636	0.0675	0.0714	0.0753
0.2	0.0793	0.0832	0.0871	0.0910	0.0948	0.0987	0.1026	0.1064	0.1103	0.1141
0.3	0.1179	0.1217	0.1255	0.1293	0.1331	0.1368	0.1406	0.1443	0.1480	0.1517
0.4	0.1554	0.1591	0.1628	0.1664	0.1700	0.1736	0.1772	0.1808	0.1844	0.1879
0.5	0.1915	0.1950	0.1985	0.2019	0.2054	0.2088	0.2123	0.2157	0.2190	0.2224
0.6	0.2257	0.2291	0.2324	0.2357	0.2389	0.2422	0.2454	0.2486	0.2517	0.2549
0.7	0.2580	0.2611	0.2642	0.2673	0.2704	0.2734	0.2764	0.2794	0.2823	0.2852
0.8	0.2881	0.2910	0.2939	0.2967	0.2995	0.3023	0.3051	0.3078	0.3106	0.3133
0.9	0.3159	0.3186	0.3212	0.3238	0.3264	0.3289	0.3315	0.3340	0.3365	0.3389
1.0	0.3413	0.3438	0.3461	0.3485	0.3508	0.3531	0.3554	0.3577	0.3599	0.3621
1.1	0.3643	0.3665	0.3686	0.3708	0.3729	0.3749	0.3770	0.3790	0.3810	0.3830
1.2	0.3849	0.3869	0.3888	0.3907	0.3925	0.3944	0.3962	0.3980	0.3997	0.4015
1.3	0.4032	0.4049	0.4066	0.4082	0.4099	0.4115	0.4131	0.4147	0.4162	0.4177
1.4	0.4192	0.4207	0.4222	0.4236	0.4251	0.4265	0.4279	0.4292	0.4306	0.4319
1.5	0.4332	0.4345	0.4357	0.4370	0.4382	0.4394	0.4406	0.4418	0.4429	0.4441
1.6	0.4452	0.4463	0.4474	0.4484	0.4495	0.4505	0.4515	0.4525	0.4535	0.4545
1.7	0.4554	0.4564	0.4573	0.4582	0.4591	0.4599	0.4608	0.4616	0.4625	0.4633
1.8	0.4641	0.4649	0.4656	0.4664	0.4671	0.4678	0.4686	0.4693	0.4699	0.4706
1.9	0.4713	0.4719	0.4726	0.4732	0.4738	0.4744	0.4750	0.4756	0.4761	0.4767
2.0	0.4772	0.4778	0.4783	0.4788	0.4793	0.4798	0.4803	0.4808	0.4812	0.4817
2.1	0.4821	0.4826	0.4830	0.4834	0.4838	0.4842	0.4846	0.4850	0.4854	0.4857
2.2	0.4861	0.4864	0.4868	0.4871	0.4875	0.4878	0.4881	0.4884	0.4887	0.4890
2.3	0.4893	0.4896	0.4898	0.4901	0.4904	0.4906	0.4909	0.4911	0.4913	0.4916
2.4	0.4918	0.4920	0.4922	0.4925	0.4927	0.4929	0.4931	0.4932	0.4934	0.4936
2.5	0.4938	0.4940	0.4941	0.4943	0.4945	0.4946	0.4948	0.4949	0.4951	0.4952
2.6	0.4953	0.4955	0.4956	0.4957	0.4959	0.4960	0.4961	0.4962	0.4963	0.4964
2.7	0.4965	0.4966	0.4967	0.4968	0.4969	0.4970	0.4971	0.4972	0.4973	0.4974
2.8	0.4974	0.4975	0.4976	0.4977	0.4977	0.4978	0.4979	0.4979	0.4980	0.4981
2.9	0.4981	0.4982	0.4982	0.4983	0.4984	0.4984	0.4985	0.4985	0.4986	0.4986
3.0	0.4987	0.4987	0.4987	0.4988	0.4988	0.4989	0.4989	0.4989	0.4990	0.4990
3.1	0.4990	0.4991	0.4991	0.4991	0.4992	0.4992	0.4992	0.4992	0.4993	0.4993
3.2	0.4993	0.4993	0.4994	0.4994	0.4994	0.4994	0.4994	0.4995	0.4995	0.4995
3.3	0.4995	0.4995	0.4995	0.4996	0.4996	0.4996	0.4996	0.4996	0.4996	0.4997
3.4	0.4997	0.4997	0.4997	0.4997	0.4997	0.4997	0.4997	0.4997	0.4997	0.4998
3.5	0.4998	0.4998	0.4998	0.4998	0.4998	0.4998	0.4998	0.4998	0.4998	0.4998

The value in the table represents the area from the mean to the corresponding z-value.

APPENDIX 3

t DISTRIBUTION

Degrees of Freedom	α						
	0.005	0.01	0.025	0.05	0.10	0.25	One-tailed
	0.01	0.02	0.05	0.10	0.20	0.50	Two-tailed
1	63.6559	31.8210	12.7062	6.3137	3.0777	1.0000	
2	9.9250	6.9645	4.3027	2.9200	1.8856	0.8165	
3	5.8408	4.5407	3.1824	2.3534	1.6377	0.7649	
4	4.6041	3.7469	2.7765	2.1318	1.5332	0.7407	
5	4.0321	3.3649	2.5706	2.0150	1.4759	0.7267	
6	3.7074	3.1427	2.4469	1.9432	1.4398	0.7176	
7	3.4995	2.9979	2.3646	1.8946	1.4149	0.7111	
8	3.3554	2.8965	2.3060	1.8595	1.3968	0.7064	
9	3.2498	2.8214	2.2622	1.8331	1.3830	0.7027	
10	3.1693	2.7638	2.2281	1.8125	1.3722	0.6998	
11	3.1058	2.7181	2.2010	1.7959	1.3634	0.6974	
12	3.0545	2.6810	2.1788	1.7823	1.3562	0.6955	
13	3.0123	2.6503	2.1604	1.7709	1.3502	0.6938	
14	2.9768	2.6245	2.1448	1.7613	1.3450	0.6924	
15	2.9467	2.6025	2.1315	1.7531	1.3406	0.6912	
16	2.9208	2.5835	2.1199	1.7459	1.3368	0.6901	
17	2.8982	2.5669	2.1098	1.7396	1.3334	0.6892	
18	2.8784	2.5524	2.1009	1.7341	1.3304	0.6884	
19	2.8609	2.5395	2.0930	1.7291	1.3277	0.6876	
20	2.8453	2.5280	2.0860	1.7247	1.3253	0.6870	
21	2.8314	2.5176	2.0796	1.7207	1.3232	0.6864	
22	2.8188	2.5083	2.0739	1.7171	1.3212	0.6858	
23	2.8073	2.4999	2.0687	1.7139	1.3195	0.6853	
24	2.7970	2.4922	2.0639	1.7109	1.3178	0.6848	
25	2.7874	2.4851	2.0595	1.7081	1.3163	0.6844	
26	2.7787	2.4786	2.0555	1.7056	1.3150	0.6840	
27	2.7707	2.4727	2.0518	1.7033	1.3137	0.6837	
28	2.7633	2.4671	2.0484	1.7011	1.3125	0.6834	
29	2.7564	2.4620	2.0452	1.6991	1.3114	0.6830	
30	2.7500	2.4573	2.0423	1.6973	1.3104	0.6828	
40	2.7045	2.4233	2.0211	1.6839	1.3031	0.6807	
60	2.6603	2.3901	2.0003	1.6706	1.2958	0.6786	
120	2.6174	2.3578	1.9799	1.6576	1.2886	0.6765	
Large df	2.5763	2.3267	1.9602	1.6450	1.2816	0.6745	

APPENDIX 4

CHI-SQUARE DISTRIBUTION

df	Area to the Right of the Critical Value									
	0.995	0.99	0.975	0.95	0.9	0.1	0.05	0.025	0.01	0.005
1	0.000039	0.0002	0.0010	0.0039	0.0158	2.7055	3.8415	5.0239	6.6349	7.8794
2	0.0100	0.0201	0.0506	0.1026	0.2107	4.6052	5.9915	7.3778	9.2104	10.5965
3	0.0717	0.1148	0.2158	0.3518	0.5844	6.2514	7.8147	9.3484	11.3449	12.8381
4	0.2070	0.2971	0.4844	0.7107	1.0636	7.7794	9.4877	11.1433	13.2767	14.8602
5	0.4118	0.5543	0.8312	1.1455	1.6103	9.2363	11.0705	12.8325	15.0863	16.7496
6	0.6757	0.8721	1.2373	1.6354	2.2041	10.6446	12.5916	14.4494	16.8119	18.5475
7	0.9893	1.2390	1.6899	2.1673	2.8331	12.0170	14.0671	16.0128	18.4753	20.2777
8	1.3444	1.6465	2.1797	2.7326	3.4895	13.3616	15.5073	17.5345	20.0902	21.9549
9	1.7349	2.0879	2.7004	3.3251	4.1682	14.6837	16.9190	19.0228	21.6660	23.5893
10	2.1558	2.5582	3.2470	3.9403	4.8652	15.9872	18.3070	20.4832	23.2093	25.1881
11	2.6032	3.0535	3.8157	4.5748	5.5778	17.2750	19.6752	21.9200	24.7250	26.7569
12	3.0738	3.5706	4.4038	5.2260	6.3038	18.5493	21.0261	23.3367	26.2170	28.2997
13	3.5650	4.1069	5.0087	5.8919	7.0415	19.8119	22.3620	24.7356	27.6882	29.8193
14	4.0747	4.6604	5.6287	6.5706	7.7895	21.0641	23.6848	26.1189	29.1412	31.3194
15	4.6009	5.2294	6.2621	7.2609	8.5468	22.3071	24.9958	27.4884	30.5780	32.8015
16	5.1422	5.8122	6.9077	7.9616	9.3122	23.5418	26.2962	28.8453	31.9999	34.2671
17	5.6973	6.4077	7.5642	8.6718	10.0852	24.7690	27.5871	30.1910	33.4087	35.7184
18	6.2648	7.0149	8.2307	9.3904	10.8649	25.9894	28.8693	31.5264	34.8052	37.1564
19	6.8439	7.6327	8.9065	10.1170	11.6509	27.2036	30.1435	32.8523	36.1908	38.5821
20	7.4338	8.2604	9.5908	10.8508	12.4426	28.4120	31.4104	34.1696	37.5663	39.9969
21	8.0336	8.8972	10.2829	11.5913	13.2396	29.6151	32.6706	35.4789	38.9322	41.4009
22	8.6427	9.5425	10.9823	12.3380	14.0415	30.8133	33.9245	36.7807	40.2894	42.7957
23	9.2604	10.1957	11.6885	13.0905	14.8480	32.0069	35.1725	38.0756	41.6383	44.1814
24	9.8862	10.8563	12.4011	13.8484	15.6587	33.1962	36.4150	39.3641	42.9798	45.5584
25	10.5196	11.5240	13.1197	14.6114	16.4734	34.3816	37.6525	40.6465	44.3140	46.9280
26	11.1602	12.1982	13.8439	15.3792	17.2919	35.5632	38.8851	41.9231	45.6416	48.2898
27	11.8077	12.8785	14.5734	16.1514	18.1139	36.7412	40.1133	43.1945	46.9628	49.6450
28	12.4613	13.5647	15.3079	16.9279	18.9392	37.9159	41.3372	44.4608	48.2782	50.9936
29	13.1211	14.2564	16.0471	17.7084	19.7677	39.0875	42.5569	45.7223	49.5878	52.3355
30	13.7867	14.9535	16.7908	18.4927	20.5992	40.2560	43.7730	46.9792	50.8922	53.6719
40	20.7066	22.1642	24.4331	26.5093	29.0505	51.8050	55.7585	59.3417	63.6908	66.7660
50	27.9908	29.7067	32.3574	34.7642	37.6886	63.1671	67.5048	71.4202	76.1538	79.4898
60	35.5344	37.4848	40.4817	43.1880	46.4589	74.3970	79.0820	83.2977	88.3794	91.9518
70	43.2753	45.4417	48.7575	51.7393	55.3289	85.5270	90.5313	95.0231	100.4251	104.2148
80	51.1719	53.5400	57.1532	60.3915	64.2778	96.5782	101.8795	106.6285	112.3288	116.3209
90	59.1963	61.7540	65.6466	69.1260	73.2911	107.5650	113.1452	118.1359	124.1162	128.2987
100	67.3275	70.0650	74.2219	77.9294	82.3581	118.4980	124.3421	129.5613	135.8069	140.1697

INDEX

Acceptance Region, 119
Accuracy, 97
Alpha Error, 113
Alternative Hypothesis, 110
Arithmetic Mean, 47
Asymmetric, 86
Bar Chart, 39
Before-and-After Experimental Design, 128
Biased Sample, 26
Bimodal Distribution, 52, 86
Box Plot, 68
Categorical data, 143
Central Limit Theorem, 93
Chi-square formula, 145
Chi-square tests, 143
Classical Probability, 74
Cluster Sample, 28
Coefficient of Determination, 160
Confidence Interval, 104
Confidence Level, 104
Contingency table, 144
Continuous Random Variable, 81
Convenience Sample, 29
Correlation Coefficient, 154
Correlation, 154
Cumulative Frequency Distribution, 36
Cumulative Relative Frequency Distribution, 36
Data, Definition of, 5
Data, Continuous, 18
Decile, 66
Dependent Variable, 157
Dependent Samples, 129
Descriptive Statistics, 3,6,8
Discrete Data, 16
Discrete Random Variable, 81
Empirical Probability, 78
Errors, 113
Ethical Considerations, 10-14
Exhaustive, 35
Expected frequency, 143
Experiment, Definition of, 5
F-test, 134
Five Number Summary, 66
Frequency, Cumulative, 36
Frequency Polygon, 40
Frequency, 35, 36
Frequency Distribution, 35
Frequency, Relative, 36
Goodness-of-Fit Test, 144
Graphs, 37-42
Histogram, 39
Historical Sample, 29
Homogeneity of Variance, 132
Hypothesis Testing, 109
Independent Samples, 129
Independent Variable, 157
Inferential Statistics, 3,6,10
Interquartile Range, 66
Interval Data, 19
Interval Estimate, 103
J-Curve, 86
Journal Form, 130
Level of Significance, 114
Lickert Scale, 21
Line Graph, 39
Line of Best Fit, 159
Lower-tailed Test, 113
Matched Pair Design, 129
Maximum, 59
Mean, 4758
Measures of Location, 59
Measures of Dispersion, 54
Measures of Central Tendency, 47
Median, 50
Midrange, 53
Minimum, 59
Mode, 51
Mutually Exclusive, 35
Nominal Data, 16
Normal Distribution, 87
Null Hypothesis, 110
Observed frequency, 143
Ogive, 41
One-dimensional categorical problem, 143
One-tailed Test, 112
Ordinal Data, 18

p-Value, 117
Parameter, Definition of, 6
Pearson's Correlation Coefficient, 154
Percentile Rank, 64
Pie Chart, 40
Point Estimate, 98
Population, Definition of, 4
Power of a Test, 115
Precision, 97
Probability Distributions, 83
Probability, 72
Probability, Definition of, 73
Quartile, 66
Random Sample, 27-28
Random Variable, 80
Range, 54
Ratio Data, 20
Regression Line, 156
Rejection Region, 119
Relative Frequency Distribution, 36
Research Design, 3,6,75,22
Sample, Convenience, 29
Sample, Definition of, 5
Sample, Cluster, 28
Sample, Historical, 29
Sample, Unbiased, 26
Sample, Systematic random, 27
Sample, Simple Random, 26
Sample, Stratified, 28
Sample, Judgment, 30
Sample, Nonprobabilistic, 24
Sample, Biased, 26
Sample, Probabilistic, 24
Samples, 23-29
Sampling, 23
Sampling Distributions, 94
Sampling Error, 30, 101
Scatter Diagram, 153
Skewed Distribution, 86
Slope of the Regression Line, 158
Standard Deviation, 56
Standard Error, 96
Standard Score, 62
Statistic, Definition of, 5
Stem-and-Leaf Display or Plot, 38
Stratified Random Sample, 28
Student t-Distribution, 106
Symmetry, 86
Systematic Random Sample, 27
Test of Independence, 148
Test Statistic, 116
Trim Mean, 53
Two-dimensional categorical problem, 143
Two-tailed Test, 112
Type II Error, 113
Type I Error, 113
Unbiased Sample, 26
Uniform Distribution, 86
Upper-tailed Test, 112
Variable, Definition of, 4
Variance, 56
Y-intercept of the Regression Line, 158
Z-Score, 86